Clinical Canine and Feline Reproduction

EVIDENCE-BASED ANSWERS

Clinical Canine and Feline Reproduction

EVIDENCE-BASED ANSWERS

Margaret V. Root Kustritz
DVM, PhD, DACT
Associate Professor
University of Minnesota College of Veterinary Medicine
St. Paul, MN

WILEY-BLACKWELL
A John Wiley & Sons, Inc., Publication

Edition first published 2010

Blackwell Publishing was acquired by John Wiley & Sons in February 2007. Blackwell's publishing program has been merged with Wiley's global Scientific, Technical, and Medical business to form Wiley-Blackwell.

Editorial Office
2121 State Avenue, Ames, Iowa 50014-8300, USA

For details of our global editorial offices, for customer services, and for information about how to apply for permission to reuse the copyright material in this book, please see our Website at www.wiley.com/wiley-blackwell.

Library of Congress Cataloging-in-Publication Data

Root Kustritz, Margaret V.
Clinical canine and feline reproduction : evidence-based answers / Margaret V. Root Kustritz.
p. ; cm.
Includes bibliographical references and index.
ISBN-13: 978-0-8138-1584-8 (alk. paper)
ISBN-10: 0-8138-1584-3 (alk. paper)
1. Dogs–Reproduction. 2. Cats–Reproduction. I. Title.
[DNLM: 1. Dogs–physiology–Handbooks. 2. Reproduction–physiology–Handbooks. 3. Cat Diseases–Handbooks. 4. Cats–physiology–Handbooks. 5. Clinical Laboratory Techniques–veterinary–Handbooks. 6. Dog Diseases–Handbooks. SF 768.2.D6 R782c 2010]
SF992.U75R66 2010
636.7′08982–dc22

2009023115

A catalog record for this book is available from the U.S. Library of Congress.

Set in 10 on 12 pt Minion by SNP Best-set Typesetter Ltd., Hong Kong
Printed in Singapore by Markono Print Media Pte Ltd

Disclaimer

1 2010

Dedication

Non nobis nomine Domine, sed nomini tuo da gloriam.

Contents

Preface, xiii

Acknowledgments, xv

Section I Canine techniques, 3

1. What is the technique for collection of a vaginal cytology specimen?, 5
2. How do I interpret vaginal cytology?, 7
3. What equipment do I need to perform vaginoscopy and how do I interpret what I see?, 13
4. What is the technique for vaginography?, 19
5. What techniques are available for pregnancy diagnosis and when are they best used?, 21
6. What is the technique for semen collection from male dogs?, 25
7. What is the technique for semen evaluation in dogs?, 29
8. What is the technique for testicular ultrasound and how do I interpret what I see?, 35
9. What is the technique for prostatic fine-needle aspirate or biopsy?, 37
10. What is the technique for testicular fine-needle aspirate and how do I interpret what I see?, 39
11. What is the technique for testicular biopsy?, 41
12. What is the technique for preparation of chilled semen in dogs?, 43
13. What is the technique for preparation of frozen semen in dogs?, 47
14. What is the technique for vaginal insemination?, 51
15. What is the technique for transcervical insemination?, 55
16. What is the technique for surgical insemination?, 57

17. What is the technique for anesthesia for Cesarean section?, 59

18. What is the technique for Cesarean section?, 63

Section II Canine reproductive physiology, 67

19. What is the normal age for puberty onset in bitches and dogs?, 69

20. What are the normal parameters for the estrous (heat) cycle in bitches?, 71

21. What are the normal parameters for semen quality in dogs?, 75

Section III Canine reproductive management, 79

22. What age is best for ovariohysterectomy of female dogs?, 81

23. When is it best to perform ovariohysterectomy of female dogs relative to heat?, 83

24. What age is best for castration of male dogs?, 85

25. Are nonsurgical alternatives available for sterilization or contraception of female or male dogs?, 87

26. What comprises recommended pre-breeding evaluation for bitches?, 91

27. What effect does hypothyroidism have on fertility?, 93

28. What are the pros and cons of treatment with antibiotics during heat in bitches?, 95

29. How do I use progesterone and other measures for breeding management?, 97

30. What drugs are unsafe to use during pregnancy in bitches?, 101

31. Are there dietary supplements proven to improve fertility in bitches?, 105

32. How can we prevent dystocia?, 107

33. What are the normal physiologic and physical changes in bitches pre-partum?, 109

34. How can I terminate pregnancy in bitches?, 111

35. Should bitches be spayed at the time of removal of a mammary mass?, 115

36. Are there dietary supplements proven to improve semen quality in male dogs?, 117

37. Can I locate retained testes without surgery?, 119

38. Is there a medical therapy to cause testes to descend into the scrotum after 8 weeks of age?, 121

Section IV Canine disease, 123

39. How do I diagnose and treat dystocia?, 125

40. How do I diagnose and treat eclampsia (hypocalcemia)?, 129

41. What are the causes of prolonged vulvar discharge after whelping in bitches?, 131

42. What is the best treatment for puppy vaginitis? Should the dog be allowed to go through one heat cycle?, 133

43. What is the best treatment for chronic vaginitis in a spayed female dog?, 135

44. How do I test for and treat canine brucellosis?, 139

45. What are the causes of vulvar discharge during pregnancy?, 143

46. What diagnostic tests can be performed to identify the cause of pregnancy loss?, 145

47. How do I diagnose and treat pyometra in bitches?, 149

48. What are the causes of agalactia in bitches and how is it best treated?, 155

49. How do I treat mastitis in a nursing bitch?, 157

50. What are the causes of persistent estrus (heat) in dogs?, 161

51. What are the causes of lack of cycling (persistent anestrus) in dogs?, 165

52. What is the diagnostic approach for infertility of a bitch?, 169

53. How do I treat benign prostatic hypertrophy?, 173

54. How do I treat prostatitis?, 177

55. What is the best treatment for recurrent paraphimosis (extrusion of the penis from the prepuce) in neutered male dogs?, 179

56. What is the diagnostic approach for infertility of a male dog?, 181

Section V Feline techniques, 185

57. What is the technique for collection of a vaginal cytology specimen?, 187

58. How do I interpret vaginal cytology?, 189

59. What techniques are available for pregnancy diagnosis and when are they best used?, 191

60. What are the techniques for semen collection from male cats?, 193

61. What is the technique for semen evaluation in cats?, 195

62. What is the technique for vaginal insemination?, 199

63. What is the technique for surgical insemination?, 201

64. What is the technique for anesthesia for Cesarean section?, 203

65. What is the technique for Cesarean section?, 207

Section VI Feline reproductive physiology, 209

66. What is the normal age for puberty onset in queens and toms?, 211

67. What are the normal parameters for the estrous (heat) cycle in queens?, 213

68. What are the normal parameters for semen quality in cats?, 215

Section VII Feline reproductive management, 217

69. What age is best for ovariohysterectomy of female cats?, 219

70. When is it best to perform ovariohysterectomy of female cats relative to heat?, 221

71. What age is best for castration of male cats?, 223

72. Are nonsurgical alternatives available for sterilization or contraception of female or male cats?, 225

73. What comprises recommended pre-breeding evaluation for queens?, 227

74. How do I use progesterone and other measures for breeding management?, 229

75. What drugs are unsafe to use during pregnancy in queens?, 231

76. How can I terminate pregnancy in queens?, 233

77. Should queens be spayed at the time of removal of a mammary mass?, 235

Section VIII Feline disease, 237

78. How do I diagnose and treat dystocia?, 239

79. What diagnostic tests can be performed to identify the cause of pregnancy loss?, 243

80. How do I diagnose and treat pyometra in queens?, 245

81. How do I treat mastitis in a nursing queen?, 249

82. What are the causes of persistent estrus (heat) in cats?, 251

83. What are the causes of lack of cycling (persistent anestrus) in cats?, 253

84. What are the causes of estrus after ovariohysterectomy of female cats?, 255

85. What are the causes of persistent male behavior after castration of male cats?, 257

86. What is the diagnostic approach for infertility of a queen?, 259

87. What is the diagnostic approach for infertility of a male cat?, 261

Section IX Pediatric techniques, 263

88. What are some of the techniques for venipuncture?, 265

89. What are some of the techniques for collection of a urine sample?, 267

90. What is the technique for placement of an intraosseous catheter?, 269

91. How do I determine safety and efficacy, and calculate correct dosage for drugs used in puppies and kittens?, 271

Section X Pediatric physical examination and management, 275

92. What are normal physical examination findings at various ages in puppies and kittens?, 277

93. What are normal values on complete blood count, chemistry profile, and urinalysis of puppies and kittens?, 281

94. How are orphan puppies and kittens best fed?, 285

95. What flea and tick control products are safe to use in puppies and kittens?, 289

Section XI Pediatric disease, 291

96. What are the causes of failure to thrive in puppies and kittens?, 293

97. What are the common causes of diarrhea in puppies and kittens?, 299

98. What are the clinical manifestations of septicemia and how is it diagnosed and treated?, 301

99. How is neonatal isoerythrolysis diagnosed, treated, and prevented in cats?, 303

100. What are the causes of stillbirths and neonatal mortality in kittens and puppies?, 305

Resources, 307

Index, 309

Preface

This text is based on the many questions I have been asked by practicing veterinarians over the years and on my own experience as a veterinarian in general practice and as a specialist. It is intended not to cover every nuance of the discipline of small animal theriogenology but rather to be a useful reference for you as you work with clients. I have tried to include citations to relevant literature whenever possible, both to justify information provided and to make that information more readily available to you. I have also included an appendix of resources to make it easy for you to find the equipment you need to do this work in practice. I hope this text answers many of the questions you encounter and am happy to help with those questions it does not answer. Feel free to contact me at rootk001@umn.edu for further assistance. Many veterinarians learn little about small animal theriogenology in veterinary college or choose not to do this work in practice. I encourage you to join the Society for Theriogenology (www.therio.org) and to ask questions to help you learn this discipline. I find small animal theriogenology fascinating and hope this text will help you have that same experience.

Acknowledgments

I thank the veterinarians who reviewed the list of questions that became the Contents, and my many colleagues who have shared their experience and knowledge with me so I could share it with others. I especially thank my mentors, Dr. Shirley Johnston and Dr. Patricia Olson, and my parents for their constant encouragement throughout my career. Finally, I thank my loving and supportive husband, Jason, and our wonderful children, Cecilia, Marie, and Monica. Yes, you may have the computer now.

Clinical Canine and Feline Reproduction

EVIDENCE-BASED ANSWERS

Section I

Canine techniques

1

What is the technique for collection of a vaginal cytology specimen?

Anatomy

The vulvar lips cover the ventral clitoral fossa and dorsal vestibule. The vestibule is angled dorsally and extends from the vulva to the urethral orifice. Cranial to the urethral orifice is the vagina.

Pre-procedure considerations

The vagina is not sterile. A non-sterile cotton-tipped applicator, moistened with saline or tap water, is used. Some describe use of a speculum to guide placement of the swab; the author finds this irritating to the bitch and cumbersome to the operator so does not advocate its use.

Procedure

Introduce the moistened swab at the dorsal-most point of the vulvar cleft and angle it upward at a 45-degree angle, rolling it between the fingers if necessary to advance it beyond the urethral papilla (Fig. 1-1). Roll the swab against the vaginal surface and pull it straight out. Roll the swab several times over a clean glass slide (Fig. 1-2). Stain the slide with a triple stain or new methylene blue.

Post-care and complications

Occasionally, the cotton comes off the applicator stick. This is easily retrieved digitally. Neither exact site in the vagina from which the sample is retrieved nor staining method alters results.

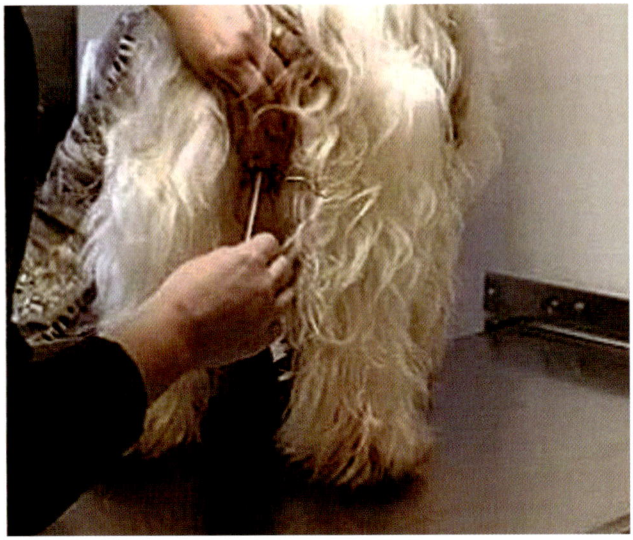

Figure 1-1: Introduction of the moistened swab at the dorsal commissure of the vulva for collection of a vaginal cytology specimen.

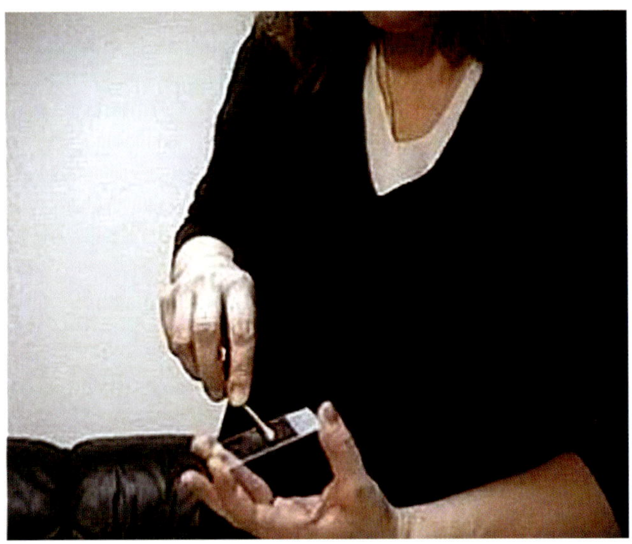

Figure 1-2: Rolling of the swab on a glass slide prior to staining.

Supplemental reading

Hiemstra M, Shaefers-Okkens AC, Teske E, et al. 2001. The reliability of vaginal cytology in determining the optimal mating time in the bitch. *Tijdschr Diergeneeskd* 126:685–689.

Root Kustritz MV. 2006. Collection of tissue and culture samples from the canine reproductive tract. *Theriogenology* 66:567–574.

2

How do I interpret vaginal cytology?

Anatomy

Not applicable.

Pre-procedure considerations

Not applicable.

Procedure

Breeding management

The healthy cuboidal epithelial cells that line the vagina at all times are termed non-cornified and include the parabasal and intermediate cells (Fig. 2-1). Under the influence of estrogen, these cells are stimulated to divide. As the cells divide and the vaginal lining thickens, the cells nearest the lumen become nonviable and lose the characteristic appearance of a healthy cell monolayer. The misshapen, clumped cells are termed cornified. Specific cell types are superficial cells and anuclear squame cells (Fig. 2-2).

During proestrus, the cell population changes from completely non-cornified in early proestrus to completely cornified in late proestrus. Red blood cells (RBCs) may be present throughout. Polymorphonuclear cells (PMNs) are present early in proestrus but are less evident as the epithelium thickens in late proestrus (Figs. 2-3, 2-4, and 2-5).

Estrus is also called standing heat. During this stage, the epithelium is at maximum thickness, with a prominent layer of keratinized cells on the luminal surface. Estrus is defined cytologically as 100% cornified cells with greater than 50% being anuclear squame cells. There are no PMNs present. RBCs may be present and bacteria are commonly evident (Fig. 2-6).

At the onset of diestrus, the cornified cells are abruptly shed. The smear is non-cornified and many PMNs may be present in the first couple of days (Fig. 2-7). Anestrus is characterized by presence of few cells, all of which are non-cornified. Occasional healthy PMNs may be seen (Fig. 2-8).

Diagnosis of disease

Spayed female dogs and bitches not in proestrus or estrus have non-cornified vaginal epithelial cells. Presence of cornified cells at any other time suggests estrogen influence. Possible disorders

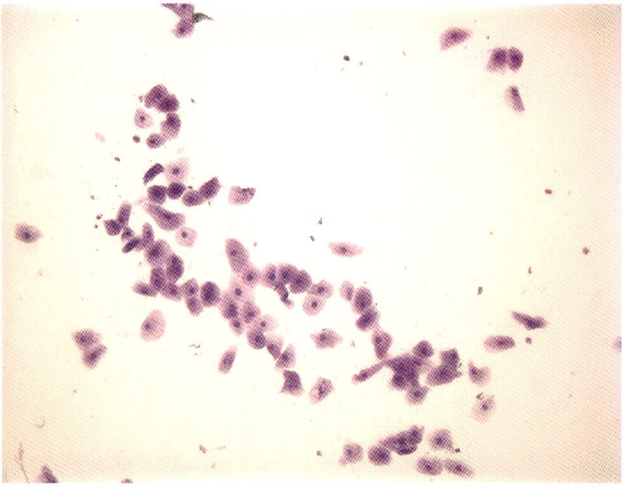

Figure 2-1: Non-cornified vaginal epithelial cells, parabasal cells (small), and intermediate cells (large).

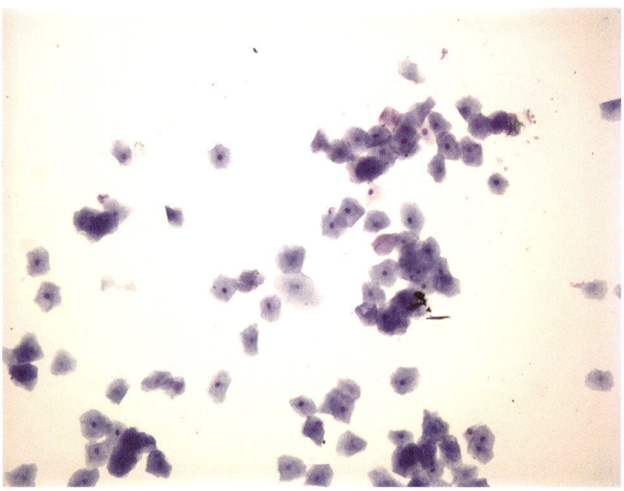

Figure 2-2: Cornified vaginal epithelial cells, superficial cells, and anuclear squame cells.

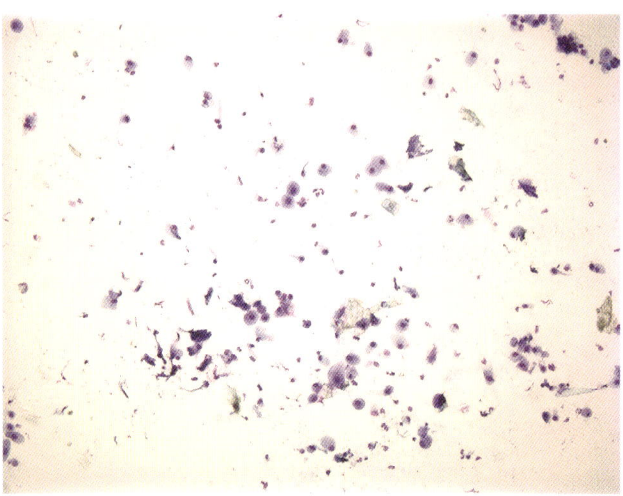

Figure 2-3: Early proestrus vaginal cytology. Note the polymorphonuclear cells.

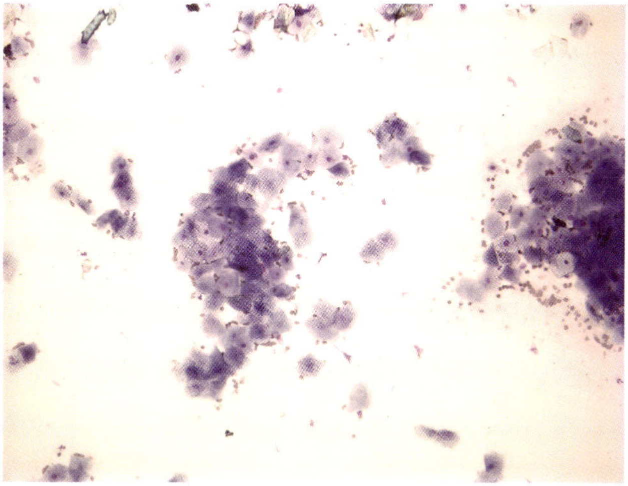

Figure 2-4: Mid-proestrus vaginal cytology.

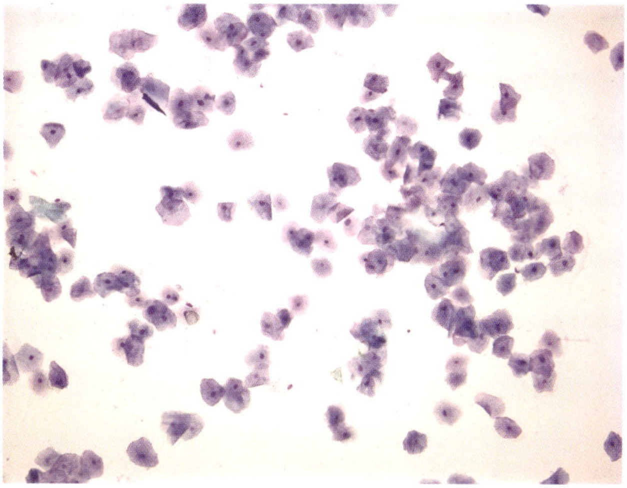

Figure 2-5: Late proestrus vaginal cytology.

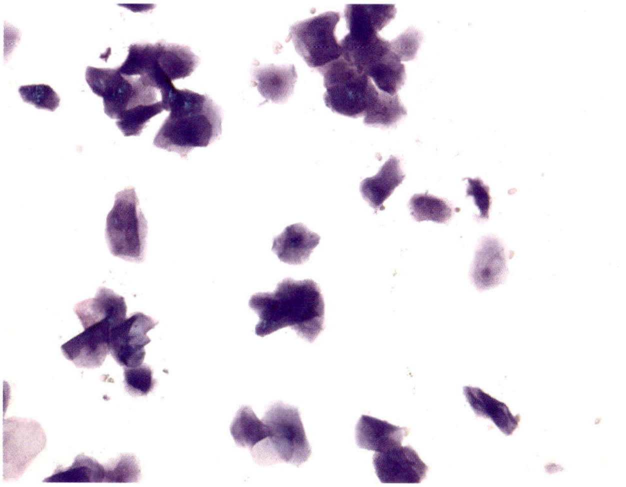

Figure 2-6: Estrus vaginal cytology. All cells are cornified and greater than 50% are anuclear squame cells. Note the bacteria.

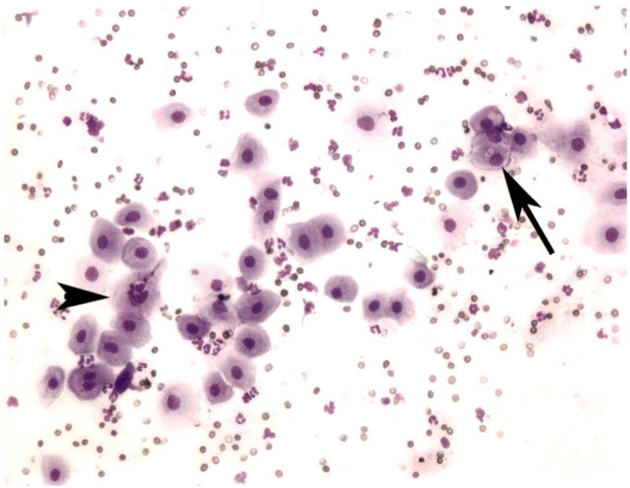

Figure 2-7: Early diestrus vaginal cytology. Note the presence of non-cornified vaginal epithelial cells and polymorphonuclear cells (PMNs). Also note the diestrum or metestrum cells (epithelial cells engulfing a PMN, arrowhead) and the foam cell (epithelial cell filled with fat vacuoles, arrow).

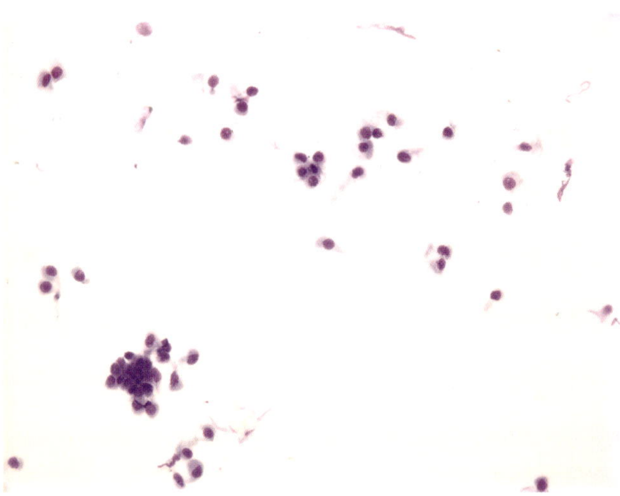

Figure 2-8: Anestrus vaginal cytology.

associated with abnormally cornified cytology include ovarian remnant syndrome, ovarian cystic disease, ovarian tumor, or estrogen ingestion or absorption, usually from a human pharmaceutical source.

Vulvar discharge may be serous, serosanguinous, hemorrhagic, mucoid, mucopurulent, or purulent. Serous discharge may be normal at the time of parturition and is not described as a common component of any disorder in dogs. Serosanguinous discharge is normal during proestrus and estrus (see Chapter 20), and postpartum, and may be evidence of abnormal estrogen secretion in dogs with ovarian remnant syndrome, ovarian cystic disease, or ovarian neoplasia. Hemorrhagic discharge is never normal and is most commonly associated with subinvolution of placental sites (see Chapter 41), neoplasia of the reproductive tract, or coagulopathy. Mucoid discharge may be evident in the first days of diestrus, as cornified cells are shed. Mucoid to purulent discharge commonly evidences inflammation of the vagina (see Chapter 43) or pyometra (see Chapter 47).

Post-care and complications

Not applicable.

Supplemental reading

Holst PA, Phemister RD. 1974. Onset of diestrus in the beagle bitch: Definition and significance. *Am J Vet Res* 35:401–406.

Johnston SD, Root Kustritz MV, Olson PN. 2001. *Canine and Feline Theriogenology*. Philadelphia: WB Saunders Co., pp. 225–242.

Rehm S, Stanislaus DJ, Williams AM. 2007. Estrous cycle-dependent histology and review of sex steroid receptor expression in dog reproductive tissues and mammary gland and associated hormonal levels. *Birth Defects Res (Part B)* 80:233–245.

Root Kustritz MV. 2008. Theriogenology question of the month: Diestrual vulvar discharge in a bitch. *J Am Vet Med Assoc* 232:841–843.

3

What equipment do I need to perform vaginoscopy and how do I interpret what I see?

Anatomy

The vulvar lips cover the ventral clitoral fossa and dorsal vestibule. The vestibule is angled dorsally and extends from the vulva to the urethral orifice. Cranial to the urethral orifice is the vagina. The cranial two-thirds of the vagina is narrowed in diameter from the presence of a tissue fold that hangs from the ceiling of the vagina, the dorsal median post-cervical fold. The cranial end of the vagina is a dead-ended fornix. The external cervical os is a rosette of tissue hanging from the ceiling of the vagina just caudal to the fornix.

Pre-procedure considerations

Vaginoscopy can be performed using either a small scope with direct lighting, such as a handheld vaginoscope (Fig. 3-1), or a long, rigid, fiber-optic endoscope (Fig. 3-2). Note that the cone of the handheld vaginoscope is not significantly different in length or diameter than the largest otoscope cone, which can be used equally successfully. The endoscope required must be very narrow in diameter; most commonly, cystoscopes, which can be passed into the urethra, are used.

Bitches in estrus tolerate vaginoscopy well. Most other bitches, including spayed female dogs, do not tolerate the procedure well. When using the handheld vaginoscope, female dogs seem to most resent the introduction of the cone. Once it has been introduced, the dog may tolerate its use. For this reason, veterinarians using this equipment may wish to attempt the technique in standing, unsedated dogs. The handheld vaginoscope can be used for an assessment of the vestibule, urethral papilla, and caudal vagina.

The long, rigid endoscope should not be used in fractious dogs because of its inflexibility and length. Female dogs should be sedated before vaginoscopy with an endoscope is performed. Place the dog in dorsal recumbency to keep her tail out of the way and to make it possible to secure her hindlimbs if necessary.

Procedure

For vaginoscopy with a handheld vaginoscope, attach a vaginoscope cone appropriate to the size of the female dog or attach the largest otoscope cone. Lubricate the cone with water-soluble lubricant. Insert the cone at the dorsal commissure of the vulva, advancing it dorsocranially. To evaluate the vagina, move the scope gently to assess the color and elasticity of the vaginal folds and to look for masses, anatomic abnormalities, discharge, lesions, or foreign objects.

Figure 3-1: Vaginoscope with stylet within cone; otoscope and cone (photo courtesy of Marie Kustritz).

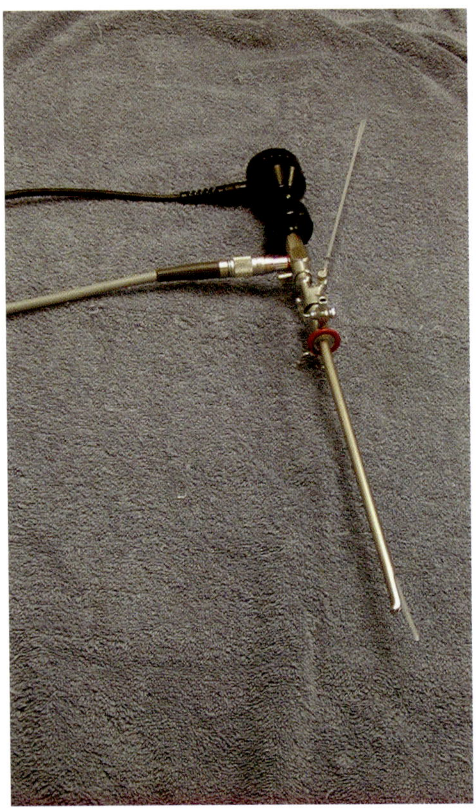

Figure 3-2: Cystoscope.

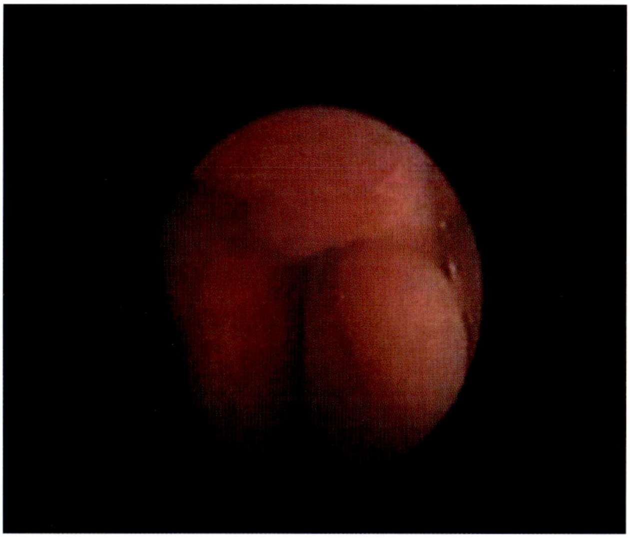

Figure 3-3: Normal vaginal mucosa.

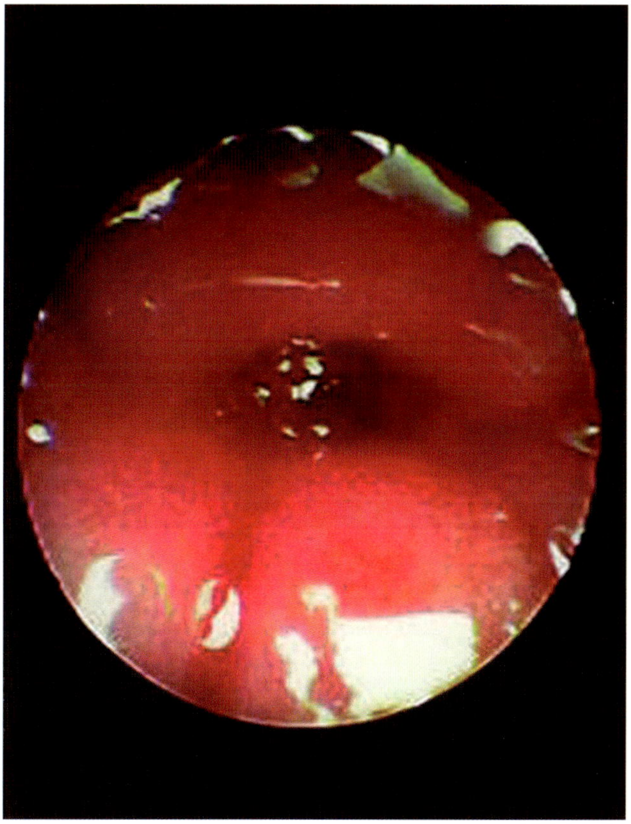

Figure 3-4: Erythematous vaginal mucosa.

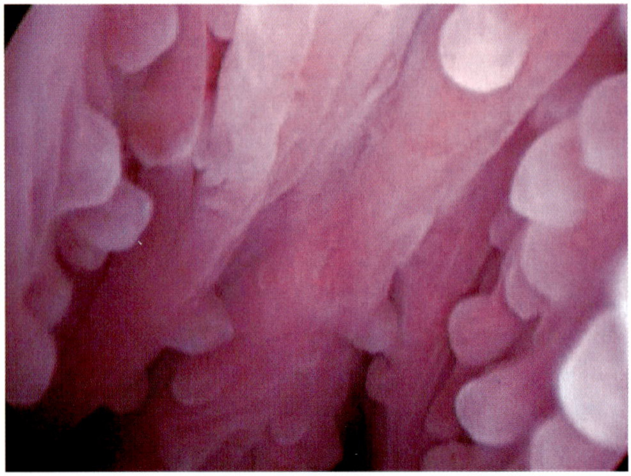

Figure 3-5: Lymphoid follicles on the vaginal mucosa (photo courtesy of Dr. Jody Lulich).

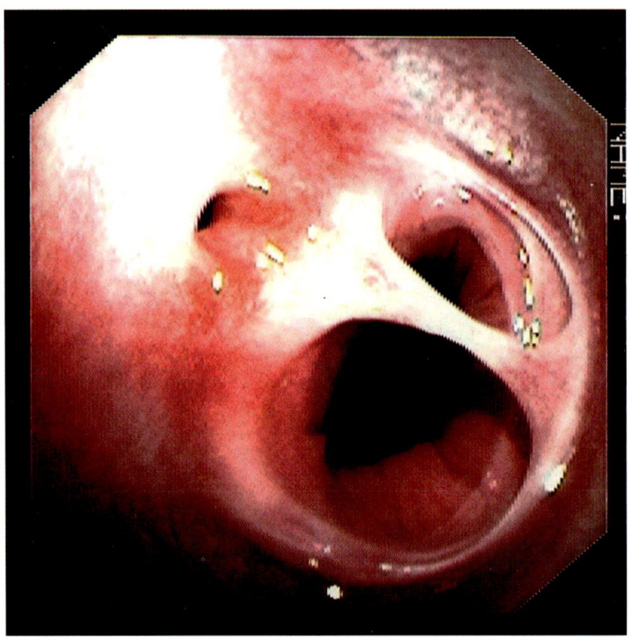

Figure 3-6: Vaginal septum (reprinted with permission from Root Kustritz MV. 2006. *The Dog Breeder's Guide to Successful Breeding and Health Management*. St. Louis, MO: Elsevier).

For vaginoscopy with a long, rigid endoscope, insert the scope as described above. Viewing is easier if the scope is attached to a camera and monitor; this may also permit capture of images for the medical record. Warmed, sterile saline may be infused through the irrigation port of the scope to ensure distension of the vaginal vault.

Normal vaginal mucosa is the healthy pink color of oral mucous membranes (Fig. 3-3). Inflamed mucosa will vary from dark pink to red (Fig. 3-4). Lymphoid follicles are nonspecific indicators of inflammation (Fig. 3-5). Other things that may be noted are discharge, urine pooling, masses or foreign objects, and anatomic abnormalities such as septa or strictures (Fig. 3-6).

Post-care and complications

Fluid will drain from the vagina after the procedure by gravity. Depending on the findings and concerns about introduction of bacteria, empirical therapy with a broad-spectrum antibiotic may be considered.

Supplemental reading

Lulich JP. 2006. Endoscopic vaginoscopy in the dog. *Theriogenology* 66:588–591.

4

What is the technique for vaginography?

Anatomy

The vulvar lips cover the ventral clitoral fossa and dorsal vestibule. The vestibule is angled dorsally and extends from the vulva to the urethral orifice. Cranial to the urethral orifice is the vagina. The cranial two-thirds of the vagina is narrowed in diameter from the presence of a tissue fold that hangs from the ceiling of the vagina, the dorsal median post-cervical fold. The cranial end of the vagina is a dead-ended fornix. The external cervical os is a rosette of tissue hanging from the ceiling of the vagina just caudal to the fornix.

Pre-procedure considerations

Vaginography requires heavy sedation or general anesthesia. Place the dog in lateral recumbency.

Procedure

Dilute iodinated contrast medium (iothalamate meglumine 60%) with an equal volume of lactated Ringer's solution. A total volume of 1 to 5 ml/kg should be prepared. Fill the lumen of a balloon-tipped catheter with contrast medium to prevent introduction of air bubbles into the vagina. Introduce the catheter into the vestibule and inflate the balloon so it sits just caudal to the urethral papilla. Infuse contrast medium into the vagina until back pressure is felt on the syringe. Obtain lateral and ventrodorsal radiographic views.

The normal vagina is bottle-shaped, with the spoon-shaped cervix visible cranially. A slight constriction, the cingulum, lies just cranial to the urethral papilla (Fig. 4-1). Vaginal anomalies will be visible as filling defects (Fig. 4-2). During proestrus and estrus, and postpartum, the cervix may be open, allowing contrast medium to move into the uterine horns.

Post-care and complications

Fluid will drain from the vagina by gravity. For dogs in whom contrast moves into the uterus, the effect of contrast medium on the endometrium and possible effects on future fertility are not defined.

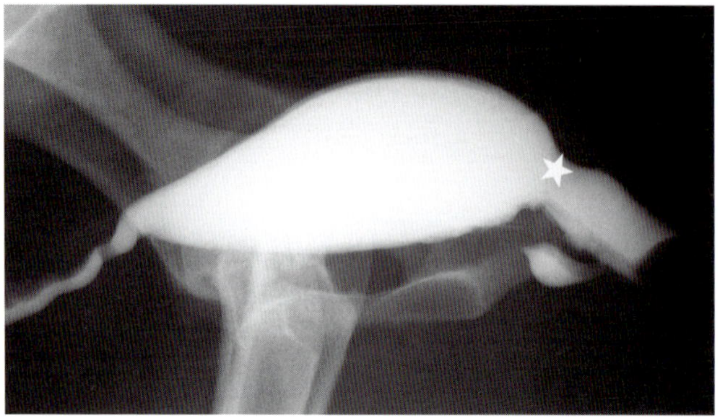

Figure 4-1: Normal vaginogram, lateral radiographic view. Note the cingulum, the normal narrowing just cranial to the urethral papilla (star).

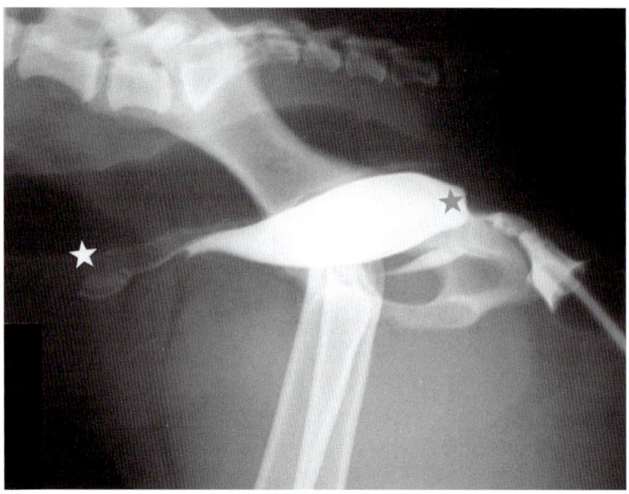

Figure 4-2: Vaginogram demonstrating circumferential stricture just cranial to the urethral papilla. Gray star denotes stricture; white star denotes spoon-shaped cervix.

Supplemental reading

Johnston SD, Root Kustritz MV, Olson PN. 2001. *Canine and Feline Theriogenology*. Philadelphia: WB Saunders Co., pp. 225–242.

5

What techniques are available for pregnancy diagnosis and when are they best used?

Anatomy

The uterine body is the only organ between the colon and the urinary bladder in the caudal abdomen. The uterine horns are tortuous and lie among the small intestines.

Pre-procedure considerations

Stage at which pregnancy can be diagnosed varies with technique used. There are few reasons to see potentially pregnant dogs outside this window and, in fact, examination may expose them to disease or otherwise potentially harm the pregnancy.

Procedure

Diagnostic tests valuable for pregnancy diagnosis in bitches and the time they are best used are listed in Table 5-1. There are no early pregnancy tests for dogs. Human early pregnancy tests do not work in dogs since they assay a hormone produced only by humans, human chorionic gonadotropin.

Post-care and complications

There are no reported detriments to the bitch or pup with any of the diagnostic techniques if performed at the proper time with technical skill and use of well-maintained equipment.

Supplemental reading

Aissi A, Slimani C. 2008. Time of initial detection of fetal structures and anatomic differentiation by using B-mode ultrasound examination in bitches. *Pak J Biol Sci* 11:1750–1753.

Kim B-S, Son C-H. 2007. Time of initial detection of fetal and extra-fetal structures by ultrasonographic examination in Miniature Schnauzer bitches. *J Vet Sci* 8:289–293.

Lenard ZM, Hopper BJ, Lester NV, et al. 2007. Accuracy of prediction of canine litter size and gestational age with ultrasound. *Aust Vet J* 85:222–225.

Purswell BJ, Parker NA, Hess M, et al. 2000. Managing pregnant and whelping dogs. *Vet Med* 95:793–800.

Root Kustritz MV. 2005. Pregnancy diagnosis and abnormalities of pregnancy in the dog. *Theriogenology* 64:755–765.

Ulutas PA, Musal B, Kiral F, et al. 2008. Acute phase protein levels in pregnancy and oestrus cycle in bitches. *Res Vet Sci* doi:10.1016/j.rvsc.2008.09.001.

Table 5-1. Pregnancy diagnostic techniques for the bitch.

Technique	Time When Best Used	Comments	Litter Size?	Viability?
Abdominal palpation	28 to 35 days of pregnancy	Prior to 28 days, the individual amniotic vesicles are difficult to feel and after 35 days, they enlarge and become confluent, again making them difficult to feel as individual entities. Care should be taken not to squeeze the amniotic vesicles but rather to let them "blip" through the fingers; vigorous palpation can injure fetuses. Palpation is difficult in obese and tense bitches.	Palpation is a poor indicator of litter size.	Palpation cannot be used to assess viability of pups. At term, puppy movement may be visible or palpable but lack of movement is not invariably associated with puppy death.
Transabdominal ultrasonography	Beyond 24 days of pregnancy	At 24 to 30 days, the amniotic vesicles are visible as black balls with a comma-shaped tissue mass within them (Fig. 5-1). Beyond 23 to 30 days, beating hearts can be seen. Fetal heart rate should be consistently greater than 200 bpm. Beyond 32 to 36 days, movement can be seen. Mineralization of the skeleton is visible beyond 35 days.	Ultrasonography is a poor indicator of litter size unless the litter is very small and the pups are widely spaced in the abdomen, and if evaluation is done prior to 41 days from ovulation. In one study, litter size was accurately predicted in only 65% of cases.	Ultrasonography is the best indicator of viability. Beyond assessment of movement, heart rate can be assessed; heart rate of less than 150 to 170 bpm is indicative of fetal stress.
Abdominal radiographs	Beyond 42 to 45 days of pregnancy	Mineralization of fetuses must be present. Radiographs are most useful very late in pregnancy, when not only can pregnancy be diagnosed but litter size, size of individual pups, and some notion of viability may also be elucidated (Fig. 5-2).	Radiography is the best indicator of litter size. Miscounts are most common in very large litters (nine or more pups).	If pups have been dead for at least 1 day, signs of fetal death may be visible on radiographs, including gas within and around the pups, and collapse of the skull or axial skeleton.

Table 5-1. *Continued*

Technique	Time When Best Used	Comments	Litter Size?	Viability?
Relaxin assay	Beyond 28 days of pregnancy	Relaxin is secreted to a much greater extent in pregnant dogs than in nonpregnant dogs in diestrus. A positive result may be evident as early as 21 days post breeding but all negatives must be rechecked no sooner than 7 days later. Relaxin stays positive for a variable time after pregnancy loss.	There is no correlation between litter size and result of the relaxin assay.	There is no correlation between viability and result of the relaxin assay.
Acute-phase proteins	Beyond 28 days of pregnancy	Acute-phase proteins are secreted in increasing amounts in the presence of physiological inflammation, such as pregnancy and placentation, or pathology, such as uterine disease. Concentrations of fibrinogen greater than 280 to 300 mg/dl are indicative of pregnancy but because pyometra occurs at the same stage of the estrous cycle as does pregnancy, measurement of fibrinogen does not differentiate these conditions. Many assays report results in g/dl, making interpretation of results difficult.	There is no correlation between litter size and results of the fibrinogen assay.	There is no correlation between fibrinogen assay results and fetal viability.

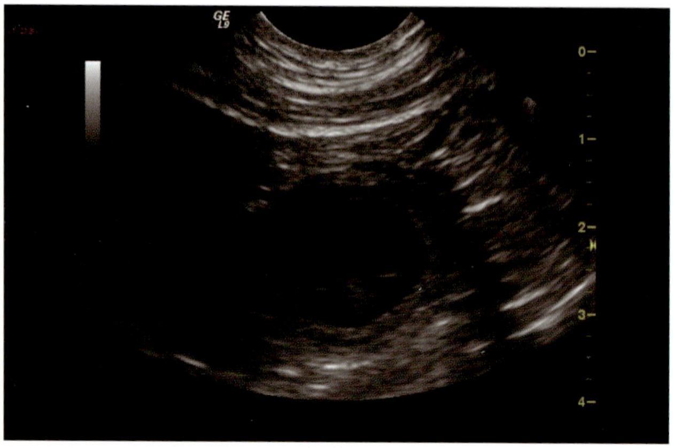

Figure 5-1: Mid-pregnancy ultrasound in a bitch. Note the fluid-filled amniotic vesicle containing an echoic embryo.

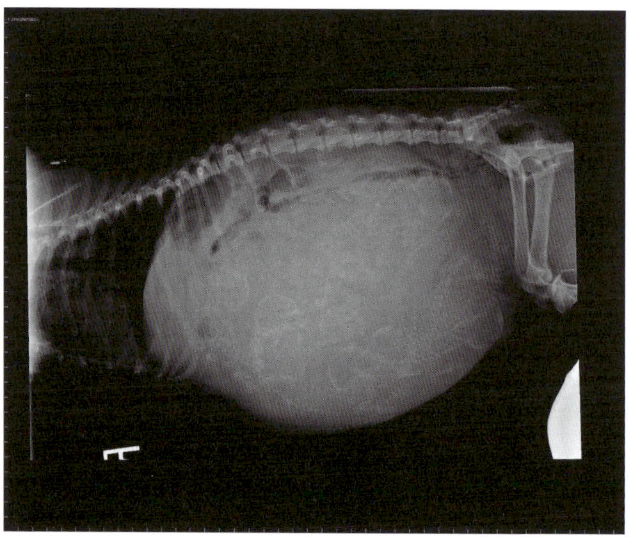

Figure 5-2: Late gestation radiograph in a bitch. Nine puppies are visible.

6

What is the technique for semen collection from male dogs?

Anatomy

The canine penis lies within the prepuce. The penis is made up of the distal glans, the pars longa glandis, which contains the os penis, and the proximal bulbus glandis, which trebles in size as the penis becomes erect. The prostate encircles the urethra at the neck of the urinary bladder and is the only accessory sex organ in dogs, providing the majority of the fluid portion of the ejaculate.

Pre-procedure considerations

Semen can be collected into any clean container. Cold shock is not as great a concern in dogs as in other species. Many operators prefer the use of some sort of a collection sleeve, or artificial vagina, made of rubber or polypropylene, with a centrifuge tube at the end as the actual collection vessel (Fig. 6-1). Sterile, disposable polypropylene collecting cones also are available (see Resources). This arrangement best mimics natural breeding as it provides circumferential pressure on the penis as would occur during intromission; permits the operator to tightly encircle the penile shaft between the bulbus glandis and the body wall with the fingers, mimicking the copulatory lock, or tie; and prevents loss or contamination of the sample.

Semen collection should be attempted in a fairly quiet area with a nonslip surface. Consistent use of one room or one rug acts as a training aid if the procedure is to be performed more than once. Presence of a teaser bitch, especially if she is in estrus, is associated with ejaculation of better quality semen. Causes of poor libido and poor semen quality are discussed in Chapter 56.

Procedure

If a teaser bitch is present, she should be restrained such that her hindquarters are on the rug, and should be muzzled if necessary to ensure that she does not endanger the dog. The male dog is presented to her, and is allowed to lick at her vulva and mount her if he so chooses. Massage the area of the bulbus glandis through the prepuce, briskly and enthusiastically. As soon as erection begins, use the hand holding the prepuce to put pressure on the caudal bulbus glandis and "squirt" the penis out of the prepuce while using the hand holding the collecting cone to push the preputial skin proximal to the engorging bulbus glandis. The goal is to make sure that the penis does not become erect within the prepuce; this is painful for some dogs. Encircle the penile

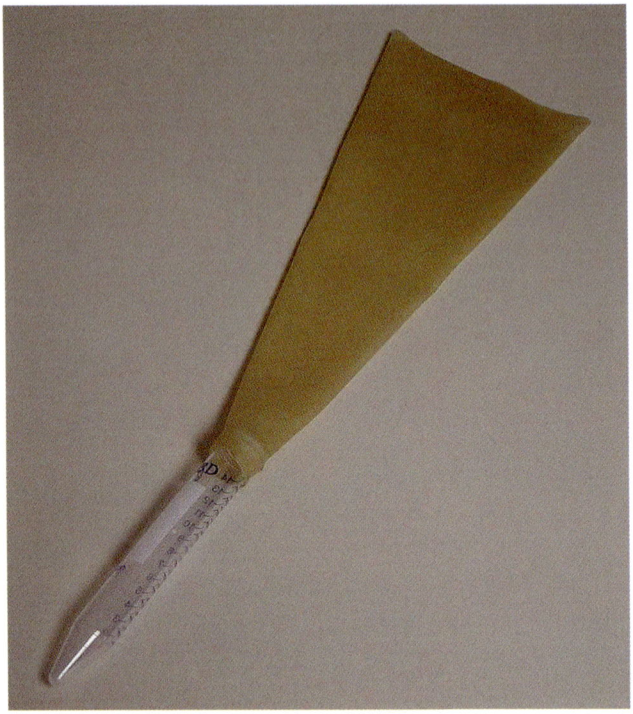

Figure 6-1: Rubber collecting cone for collection of canine semen (reprinted with permission from Root Kustritz MV. 2006. *The Dog Breeder's Guide to Successful Breeding and Health Management*. St. Louis, MO: Elsevier).

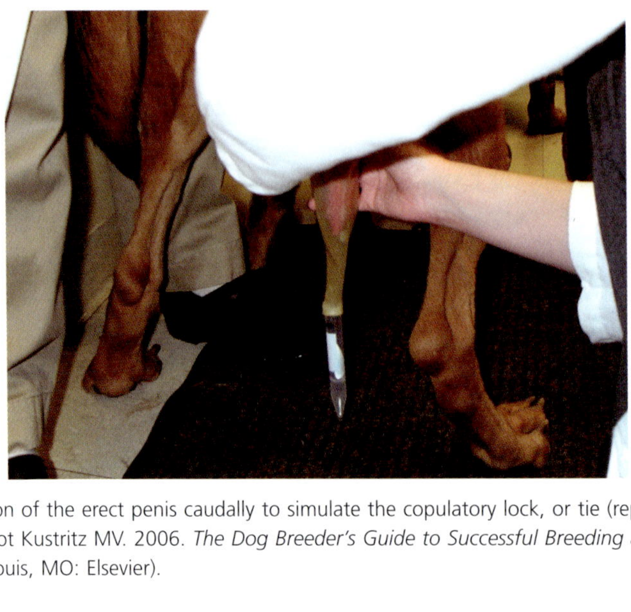

Figure 6-2: Direction of the erect penis caudally to simulate the copulatory lock, or tie (reprinted with permission from Root Kustritz MV. 2006. *The Dog Breeder's Guide to Successful Breeding and Health Management*. St. Louis, MO: Elsevier).

shaft tightly through the collecting cone just proximal to the bulbus glandis. You cannot constrict it enough with your hand to stop ejaculation.

The dog ejaculates three fractions. The first fraction is ejaculated as the dog thrusts vigorously. This fraction is of fairly small volume and is clear; it is prostatic in origin. The second fraction may be ejaculated while the dog is thrusting or just after he stops the vigorous thrusting behavior. Many dogs attempt to lift a leg to form the tie after thrusting is done. The operator should pick up the dog's leg and move the penis in a horizontal plane until it is directed caudally (Fig. 6-2). The second fraction is the sperm-rich fraction, and is white. The third fraction is prostatic in origin and is secreted in pulses that will be palpable in the hand holding the collecting cone. Concurrent anal contractions are evident. Once the third fraction is present, no more sperm-rich fluid will be collected.

Post-care and complications

Some inexperienced dogs will exhibit bleeding from small vessels on the penile surface with complete erection. This is not dangerous to the dog and usually is not a persistent problem. Some dogs will have persistent erection with prolonged secretion of prostatic fluid. Ways to encourage detumescence of the penis include walking the dog away from the teaser bitch and the collection area; gently cold-packing the dog's penis with cool cloths or providing cool hydrotherapy in a tub; or feeding the dog. Make sure complete detumescence has occurred and that the tip of the penis is not protruding before kenneling the male dog.

Supplemental reading

Kutzler MA. 2005. Semen collection in the dog. *Theriogenology* 64:747–754.

7

What is the technique for semen evaluation in dogs?

Anatomy
Not applicable.

Pre-procedure considerations
Cold shock is not a problem in dogs as it is in other species. Equipment and samples can be maintained at room temperature for evaluation.

Procedure
Color
Color is assessed visually. Normal semen is milky white (Fig. 7-1). Abnormal colors that may be seen include clear (no spermatozoa in the ejaculate), yellow (urine contamination), brown (old blood from prostate disease), red (blood from penile trauma or prostate disease), or green (infection).

Volume
Volume varies depending on how much of the third, or prostatic, fraction of the ejaculate was collected by the operator. Volume is not correlated with quality. Record volume collected before any samples are removed for evaluation; this value will be needed later to calculate total number of spermatozoa in the ejaculate.

Motility
Motility should be assessed soon after semen collection. Place one drop of semen on a glass slide. You may or may not use a cover slip. Subjectively assess the percentage of spermatozoa that are moving forward; normal is 70% or greater. Some people also assess the speed of movement of the spermatozoa; there are no reported correlations between speed of motility and fertility in that dog or with use of that particular sample.

Concentration/total number
Make a 1 : 100 dilution of semen by diluting 1 part semen (0.1 ml) with 9 parts formal-buffered saline (0.9 ml) to make a 1 : 10 dilution and then mixing 1 part of that initial dilution with 9 parts

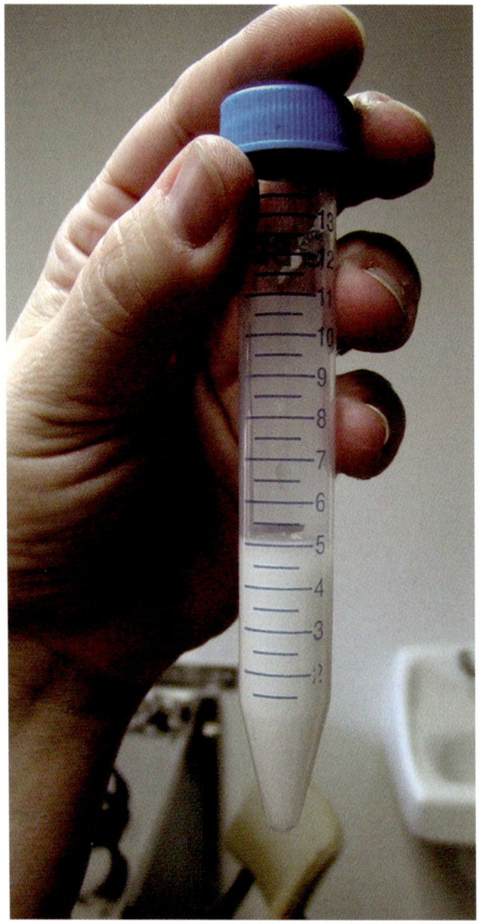

Figure 7-1: Normal semen color. Volume measurement can be read directly off the markings on the centrifuge tube (reprinted with permission from Root Kustritz MV. 2006. *The Dog Breeder's Guide to Successful Breeding and Health Management*. St. Louis, MO: Elsevier).

formal-buffered saline. This can also be done using the Unopette system for white blood cell counts (Fig. 7-2). Use the enclosed piercing device to make a hole on the top of the diluent container. Remove the piercing device to reveal the pipette. Fill the pipette by capillary action. Squeeze the diluent container and insert the pipette, letting go of everything at once so the semen is sucked from the pipette into the diluent. Remove the pipette and turn it around, reseating the hub to form a dispenser.

With either technique, spermatozoa are counted using a hemacytometer. Place the glass cover slip over the central area of the hemacytometer. Dispense diluted semen such that capillary action carries it across the central area. Fill each side of the hemacytometer separately. Allow the hemacytometer to sit for about 5 min after filling, to allow spermatozoa to settle.

The hemacytometer grid consists of nine large squares. Using the 10X objective, one of these large squares fills the microscope field. Count all the spermatozoa visible in one of the nine large squares. This yields the concentration in millions per milliliter.

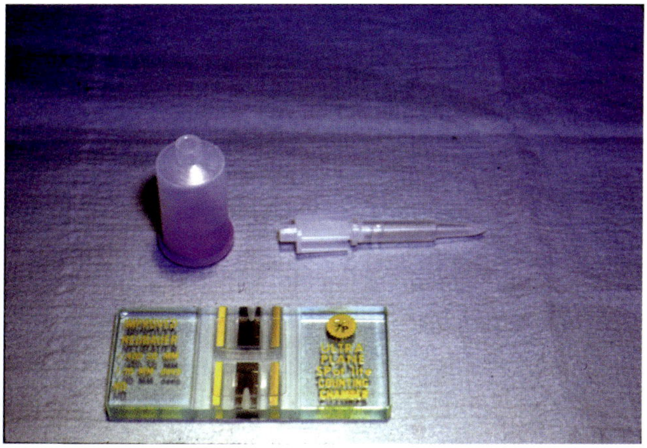

Figure 7-2: Unopette system and hemacytometer for determination of concentration of spermatozoa in canine semen (reprinted with permission from Root Kustritz MV. 2006. *The Dog Breeder's Guide to Successful Breeding and Health Management*. St. Louis, MO: Elsevier).

Other, less time-consuming, techniques for evaluation of number of spermatozoa in the ejaculate have been assessed. None, including computer-assisted systems, have been demonstrated to be as accurate as the hemacytometer method.

Concentration varies with semen collection technique; if much prostatic fluid was collected, the sample will be dilute with low concentration of spermatozoa. However, the total number of spermatozoa in the sample does not vary with technique and is the valued number in semen evaluation. Volume (milliliter per ejaculate) multiplied by concentration (spermatozoa per milliliter) yields total number (spermatozoa per ejaculate). Normal value is dependent on the size of the dog, ranging from 300 million to 2 billion.

Morphology

Morphology, or shape, of spermatozoa is not specifically associated with fertility in dogs as in other species. However, a higher percentage of morphologically normal spermatozoa is desirable as poor morphology is often associated with poor motility and, presumably, with a decreased ability to reach the uterine tube and fertilize ova. There is also some reported correlation between abnormalities of head shape and degree of DNA denaturation within the cell.

Place one drop of semen at one end of a glass slide and smear it out as for a blood smear. Let it air-dry and stain with a triple stain, leaving it in each of the three solutions (fixative, safranin, and crystal violet) for 5 min. Rinse and allow to air-dry before evaluating under oil immersion.

Another technique is to place one drop of semen at one end of a glass slide and place a drop of an eosin-nigrosin stain next to it. Using a pusher slide, rock the two drops together to mix them and draw it out into a thick smear. Allow to air-dry and evaluate under oil immersion.

Morphology can be broken down into normal spermatozoa, those with primary defects, and those with secondary defects (Fig. 7-3). Primary defects occur during spermatogenesis and include anything doubled, any abnormality of the head shape, proximal cytoplasmic droplets, and bending of the midpiece. Secondary defects occur during sample preparation or are an indication of infection and include detached heads, distal cytoplasmic droplets, and bent tails. Normal is 80% or greater morphologically normal spermatozoa. Significance of specific primary or secondary defects is not known in dogs, although a preponderance of primary defects may

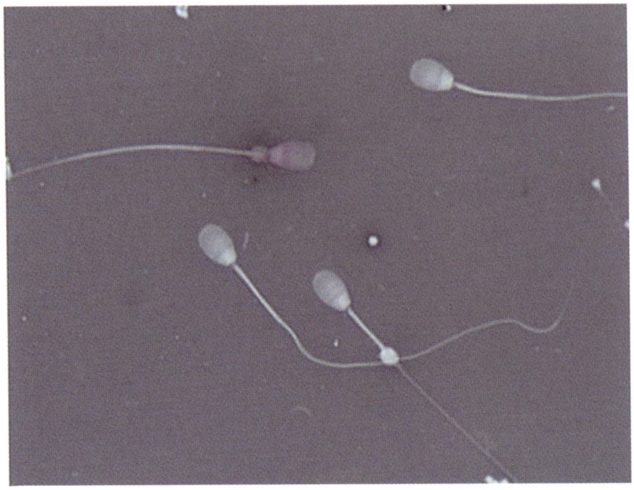

Figure 7-3: Two normal spermatozoa: one spermatozoon with a primary cytoplasmic droplet (primary defect), one spermatozoon with a distal cytoplasmic droplet (secondary defect).

carry a worse prognosis than a preponderance of secondary defects, simply because the underlying cause is less likely to be something easily addressed.

Miscellaneous tests

Other tests that may be indicated, based on clinical presentation of the animal and other semen evaluation findings, are cytology and culture. The dilute nature of canine semen often makes examination of raw specimens for cytology less than useful. Centrifuge the sample and example the pellet for inflammatory cells, abnormal epithelial cells, bacteria, or other cell types. Cytology is not well associated with culture results; if infection is suspected, culture should be performed even if cytology is not inflammatory. Because semen collection is not a sterile procedure, culture results must be interpreted recognizing the presence of normal flora. Growth of any single organism at 10^5 bacteria/ml or greater is considered significant.

In human medicine, tests for the fertilizing capability of spermatozoa are routinely performed. These tests are not commonly used in veterinary medicine. One simple test that could be used to assess the integrity of the sperm plasma membrane is the hypo-osmotic swelling test. In this test, spermatozoa are placed in a hypo-osmotic medium. If the plasma membranes are intact, suggesting that the spermaotozoa are normal and viable, fluid will leave the spermatozoa cells and the tail will curl. One simple protocol involves placement of 10 µl of semen into 100 µl of a 100 mM sucrose solution, with assessment of sperm morphology after 1 min.

Use of automated semen evaluation systems

Many facilities that do many semen evaluations, especially those that freeze semen, use an automated system for semen evaluation. These are commonly called computer-assisted sperm analysis (CASA) systems. CASA systems must be specifically calibrated for dog semen, which is more dilute than that of other species.

Post-care and complications

Not applicable.

Supplemental reading

Christensen P, Stryhn H, Hansen C. 2005. Discrepancies in the determination of sperm concentration using Burker-Turk, Thoma and Makler counting chambers. *Theriogenology* 63:992–1003.

Nunez-Martinez I, Moran JM, Pena FJ. 2005. Do computer-assisted, morphometric-derived sperm characteristics reflect DNA status in canine spermatozoa? *Reprod Dom Anim* 40:537–543.

Pinto CRF, Kozink DM. 2008. Simplified hypo-osmotic swelling test (HOST) of fresh and frozen-thawed canine spermatozoa. *Anim Reprod Sci* 104:450–455.

Rijsselaere T, Van Soom A, Tanghe S, et al. 2005. New techniques for the assessment of canine semen quality: A review. *Theriogenology* 64:706–719.

Root Kustritz MV. 2007. The value of canine semen evaluation for practitioners. *Theriogenology* 68:329–337.

Root Kustritz MV, Johnston SD, Olson PN, et al. 2005. Relationship between inflammatory cytology of canine seminal fluid and significant aerobic, anaerobic, or mycoplasma cultures of canine seminal fluid: 95 cases (1987–2000). *Theriogenology* 64:1333–1339.

Root Kustritz MV, Kilty C, Vollmer M. 2007. Spermatocrit as a measure of concentration of canine spermatozoa. *Vet Rec* 161:566–567.

8

What is the technique for testicular ultrasound and how do I interpret what I see?

Anatomy

The testicular tissue is tightly packed within a firm capsule, the tunica albuginea, which is encased within the parietal and visceral vaginal tunics and the spermatic fascia. The testes lie obliquely in the scrotum, with the head of the tightly adhered epididymis on the cranial end of the testis and the tail of the epididymis at the caudal end of the testis. The spermatic cord contains the cremaster muscle, ductus deferens, and testicular artery and veins. Within the testis, the individual seminiferous tubules empty into a central, fibrous rete testis.

Pre-procedure considerations

Most dogs tolerate testicular ultrasound without sedation. Sedation may be required if manipulation of the testes or scrotum is painful to the dog. Even dogs with thick haircoat usually have minimal haired skin on the scrotum. Scrotal skin is sensitive and elevations in intrascrotal temperature from dermatitis might be associated with decreased spermatogenesis or damage to testicular tissue. For these reasons, the scrotum is not shaved.

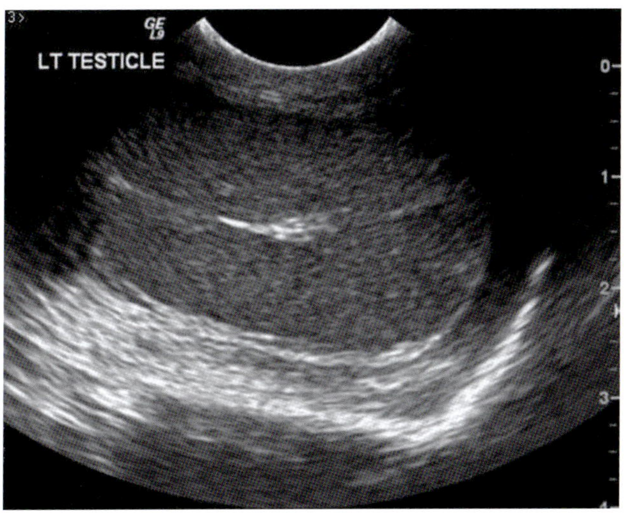

Figure 8-1: Ultrasonogram of a normal canine testis.

Procedure

Apply ultrasound gel. A transducer ranging from 4 to 7 MHz may be used to assess testicular tissue. The hyperechoic rete testis should be visible in the center of each testis. The rest of the testicular parenchyma should be uniformly dappled (Fig. 8-1). Skilled operators may be able to identify the fluid-filled epididymis. There should be no free fluid in the scrotum.

Post-care and complications

Abnormalities identified may be diagnosed by fine-needle aspirate (see Chapter 9), or histopathology after removal of a biopsy specimen (see Chapter 10) or the entire testis. Testicular ultrasound is a more accurate indicator of true testicular volume than is external measurement of testicular size with a pair of calipers.

Supplemental reading

Gouletsou PG, Galatos AD, Leontides LS. 2008. Comparison between ultrasonographic and caliper measurements of testicular volume in the dog. *Anim Repro Sci* 108:1–12.

Johnston SD, Root Kustritz MV, Olson PN. 2001. *Canine and Feline Theriogenology*. Philadelphia: WB Saunders Co., pp. 312–322.

9

What is the technique for prostatic fine-needle aspirate or biopsy?

Anatomy

The prostate encircles the urethra at the neck of the urinary bladder. Depending on its size, it may lie within the pelvis or within the abdominal cavity. The prostatic parenchyma is broken into lobules separated by connective tissue and is enclosed within a tight capsule.

Pre-procedure considerations

Fine-needle aspirate can usually be performed on unsedated animals. Biopsy requires sedation as it is a more invasive and painful procedure. Blind techniques have been reported in the veterinary literature but are inherently inferior to ultrasound-guided techniques, which permit collection of samples from areas of interest and minimize risks associated with perforation of fluid-filled areas of the gland, including the urethra. Only ultrasound-guided techniques are described.

Procedure

Sedate the animal if necessary and place it in dorsal recumbency for ultrasound. Shave the groin of hair. Identify the hypoechoic urinary bladder and look for the prostate caudal to the main body of the urinary bladder. The tight prostatic capsule is hyperechoic. The urethra is visible in the center of the prostate as a hypoechoic round or V-shaped area. Normal prostatic parenchyma is lightly mottled and uniform. Specific abnormalities are described in Chapters 53 and 54.

For fine-needle aspirate, prep the skin as for surgery. Cover the ultrasound probe with a sterile sleeve. Attach a sterile, long, 22-gauge needle to a sterile 6 cc syringe. Pass the needle through the port on the side of the transducer or alongside the transducer, watching it pass through the parenchyma to the area of interest. Exert negative pressure several times, repositioning the needle as necessary. Release pressure and withdraw the needle. Expel the contents onto a glass slide for cytology assessment or into a transport medium for culture. Cytology should be assessed by a pathologist or an experienced cytologist.

For biopsy, prep the skin as for surgery. Cover the ultrasound probe with a sterile sleeve. While viewing the prostate, pass a triggered biopsy instrument into the prostate. Trigger the instrument and retrieve the sample. Make sure the sample appears adequate for histopathology and/or culture before reversing sedation. Histopathology should be assessed by a pathologist.

Post-care and complications

Transient hematuria (no more than 4 days duration) is a reported complication of prostatic fine-needle aspirate and biopsy. Transient hemospermia may also occur. In human medicine, prophylactic antibiotic therapy is recommended after prostatic biopsy; this may be of less importance in dogs because of variations in biopsy technique between species.

Supplemental reading

Bootsma AMJ, Pes MPL, Geerlings SE, et al. 2008. Antibiotic prophylaxis in urologic procedures: A systematic review. *Euro Urol* 54:1270–1286.

Root Kustritz MV. 2006. Collection of tissue and culture samples from the canine reproductive tract. *Theriogenology* 66:567–574.

10

What is the technique for testicular fine-needle aspirate and how do I interpret what I see?

Anatomy

The testicular tissue is tightly packed within a firm capsule, the tunica albuginea, which is encased within the parietal and visceral vaginal tunics, and the spermatic fascia. The testes lie obliquely in the scrotum, with the head of the tightly adhered epididymis on the cranial end of the testis and the tail of the epididymis at the caudal end of the testis. The spermatic cord contains the cremaster muscle, ductus deferens, and testicular artery and veins.

Pre-procedure considerations

Dogs in extreme pain (e.g., those with orchitis or torsion of the spermatic cord) may not require sedation for fine-needle aspirate of the testis to be performed. However, most dogs are exquisitely sensitive to manipulation of the testes and require sedation for this procedure. Clean the scrotal skin as for any fine-needle aspirate; do not shave the scrotal skin.

Procedure

The procedure may be performed with or without ultrasound guidance. Attach a sterile, 20-gauge needle to a sterile 12 cc or larger syringe. Introduce the needle on the midline of the testis, redirecting the needle and exerting negative pressure three or four times. Release pressure and withdraw the needle. Expel the contents onto a glass slide. Ensure adequacy of the sample collected before reversing sedation.

This technique does not allow evaluation of the normal testicular architecture but permits gross assessment of spermatogenesis (whether or not maturing spermatozoa are present) and assessment of infection (whether or not inflammatory cells are present). It is reported that Sertoli cells and maturing spermatozoa are readily seen but that Leydig (interstitial) cells are not identified. Polymorphonuclear cells and bacteria are seen in dogs with orchitis, and atypical spermatozoal cells and Sertoli cells in dogs with testicular neoplasia. The slide should be evaluated by a pathologist or an experienced cytologist.

Post-care and complications

Reported complications are short-term and include mild scrotal swelling and erythema; these signs should subside within 3 days. No long-term effect on libido or semen quality is apparent in the vast majority of dogs.

Supplemental reading

Dahlbom M, Makinen A, Suominen J. 1997. Testicular fine needle aspiration cytology as a diagnostic tool in dog infertility. *J Small Anim Pract* 38:506–512.

Romagnoli S, Bonaccini P, Stelletta C, et al. 2008. Clinical use of testicular fine needle aspiration (FNA) cytology in oligozoospermic and azoospermic dogs. Proceedings, International Symposium on Canine and Feline Reproduction, Vienna, Austria.

Root Kustritz MV. 2006. Collection of tissue and culture samples from the canine reproductive tract. *Theriogenology* 66:567–574.

11

What is the technique for testicular biopsy?

Anatomy

The testicular tissue is tightly packed within a firm capsule, the tunica albuginea, which is encased within the parietal and visceral vaginal tunics, and the spermatic fascia. The testes lie obliquely in the scrotum, with the head of the tightly adhered epididymis on the cranial end of the testis and the tail of the epididymis at the caudal end of the testis. The spermatic cord contains the cremaster muscle, ductus deferens, and testicular artery and veins.

Pre-procedure considerations

The three most prevalent cell types in the canine testis are the spermatogonia, or germ cells, which divide to form spermatozoa; the Sertoli cells, which support the developing spermatozoa; and the interstitial or Leydig cells, which produce testosterone. The testis is made up of seminiferous tubules. Within a cross-section of a seminiferous tubule, spermatogonia lie against the basement membrane, with Sertoli cells surrounding increasingly mature spermatozoa toward the center. The individual seminiferous tubules empty into a central, fibrous rete testis.

Testicular biopsy rarely yields information that leads to resolution of a clinical abnormality. Rather, testicular biopsy answers basic questions regarding pathologic processes within the testes (inflammation, neoplasia, degeneration) and informs the client and veterinarian regarding possible future fertility; if there are no spermatogonia present, future spermatogenesis is impossible. Testicular biopsy is reported to be less useful if overt testicular atrophy is present, evidenced by small size and soft consistency of the testes.

Procedure

Place the dog under general anesthesia. Shave and surgically prep the pre-scrotal area as for a castration. Make a single pre-scrotal incision and advance one testis to the incision. Cut through the spermatic fascia and vaginal tunics to expose the testis within the tunica albuginea. To retrieve a biopsy sample, several techniques are described. One is to stab through the testicular capsule with a small scalpel blade. A portion of the testis will well through the capsular break (Fig. 11-1). Gently shave off this piece of testicular tissue. An alternative is to make an incision through the testicular capsule through which a triggered biopsy instrument can be introduced.

Immediately place tissue samples in a fixative and submit for assessment by a pathologist. Zenker's and Bouin's fixatives are preferred to formalin, which may introduce artifactual

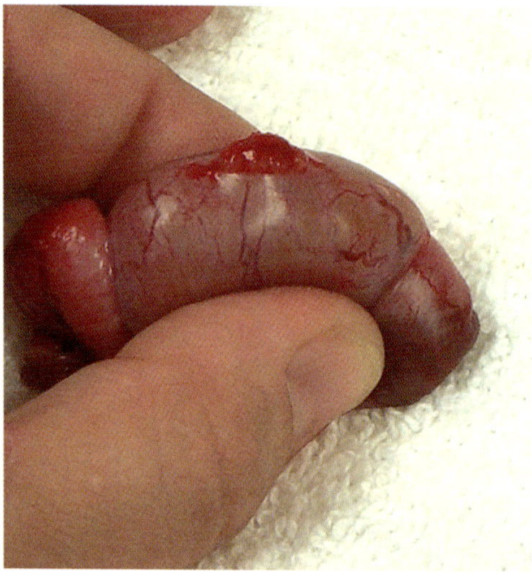

Figure 11-1: Testis excised at the time of castration to demonstrate the appearance of the welling of testicular tissue through the capsule.

defects. Close the testicular capsule with an absorbable suture. Close the skin in a routine manner.

Post-care and complications

Incision care is as for castration. Particular attention should be paid to minimizing scrotal swelling and subsequent increase in scrotal temperature, especially in dogs with hope of continuing fertility. Gross and histologic lesions were identified experimentally in dogs castrated after testicular biopsy. Gross lesions included hemorrhage and adhesion formation, seen in 31 and 23% of dogs, respectively. The most severe microscopic lesion seen was arrest in continuing spermatogenesis, noted in 26% of dogs after incisional biopsy and in only 2.6% of dogs after biopsy with a triggered instrument.

Supplemental reading

Lopate C, Threlfall WR, Rosol TJ. 1989. Histopathologic and gross effects of testicular biopsy in the dog. *Theriogenology* 32:585–602.

Root Kustritz MV. 2006. Collection of tissue and culture samples from the canine reproductive tract. *Theriogenology* 66:567–574.

12

What is the technique for preparation of chilled semen in dogs?

Anatomy

Semen is collected as for any other procedure (see Chapter 6).

Pre-procedure considerations

All male dogs used for breeding should be certified free of heritable conditions and infectious disease. All dogs should be tested and proven negative for canine brucellosis before semen is collected for chilling and shipment (see Chapter 44).

Specific tests for hereditary conditions vary by breed. Most breeds require Orthopedic Foundation for Animals or University of Pennsylvania Hip Improvement Program certification for hip dysplasia and Canine Eye Registry Foundation evaluation for ocular defects. Breeders should be referred to their national breed club web site or printed materials to see which tests are recommended by the health organization unique to their breed.

If the pups are to be registered with the American Kennel Club (AKC), the male must have a DNA profile on record with the AKC. This rule has been in effect since October 1, 1998. The DNA sample is collected using a cheek swab and may be done by the dog's owner. Breeders are referred to the AKC web site for complete information about sample collection, sample submission, and registration and payment (see Resources).

Chilled semen may be inseminated into the vagina (see Chapter 14) or the uterus (see Chapters 15 and 16). Reported conception rates with vaginal insemination average 47% and with transcervical intrauterine insemination 81%.

Procedure

Chilled semen is made up of ejaculated spermatozoa suspended in an extender. Extender is a fluid medium containing nutrients, buffers, and anti-microbials. Recipes for homemade extenders exist. Most veterinarians use a commercial extender. Extenders specifically for dogs are available commercially through several companies (see Resources). Extenders produced commercially for other species may work in dogs. In an emergency, skim milk that has been heated to 92 to 95 F for 10 min and then cooled to room temperature will work but it does not contain buffers or anti-microbials.

Collect semen in a routine fashion and maintain it at room temperature. A complete semen evaluation may be performed (see Chapter 7); at the very least, one drop should be evaluated for subjective assessment of the percentage of progressively motile spermatozoa and a superficial evaluation of concentration or total number of spermatozoa in the sample. Poor-quality semen may benefit from passage through a filter. With this process, semen is layered over a gradient and centrifuged. Abnormal spermatozoa and blood products are retained and normal spermatozoa gather into a usable pellet. If this technique is used, the centrifugation step described below is skipped. Blood in semen has no effect on viability after extension, chilling, and shipment as long as the relative volume of blood to semen is less than 10%.

If the total number of spermatozoa in a single ejaculate is low, several samples can be pooled and processed together. Techniques to increase number of spermatozoa in the ejaculate include serial collections at 30- to 60-min intervals; presence of an estrous teaser bitch; and administration of 0.1 mg/kg prostaglandin F2alpha (Lutalyse™) 15 min prior to a collection attempt. Centrifuge the semen at the setting used for urine sedimentation (300 to 500 g) for about 5 min. Draw up one drop of supernatant and evaluate it for spermatozoa; if many motile spermatozoa are still present in the supernatant, centrifuge the semen for another 5 min. Once centrifugation is complete, pour off the supernatant and resuspend the pellet of semen in a ratio of 1 part semen to 3 to 5 parts of room temperature extender. Evaluate one drop of the extended semen, again evaluating the percentage of progressively motile spermatozoa. Some companies send along specific instructions and paperwork asking for assessment of the speed of progression of spermatozoa, the percentage of morphologically normal spermatozoa, and the total number of spermatozoa in the extended sample.

Commercial companies provide shipping receptacles for the extended semen. The tube of extended semen should be tightly capped and placed in bag to ensure that the semen is not lost if the tube breaks during transport. Place the tube in a small Styrofoam box or wrap it well in newspaper, cotton, or some other cushioning product. Wrap it in a layer of newspaper. In commercial products, this is usually sandwiched between frozen ice packs in a Styrofoam or plastic enclosure for shipment. Some people report success plunging the wrapped tube of semen into ice in a thermos-type container. It has been demonstrated that thermos containers, Styrofoam boxes, and more elaborate shipping containers (see Resources) maintain semen quality equally well for up to 24 h; motility of spermatozoa remains higher beyond that time in more elaborate systems.

Shipment should ensure that the sample reaches the bitch within 24 h after collection. If properly packaged, chilled, and maintained, the sample should maintain its quality for up to 48 h and perhaps longer. Samples delayed in shipment benefit from centrifugation and suspension of the subsequent pellet in new extender prior to insemination. Packaged samples can be shipped counter-to-counter at the airport or a commercial shipper can be used. The advantage of using airlines is that the sample can be collected shortly before a flight is due to leave and the bitch inseminated soon after its arrival, permitting a very short turnaround time from collection to insemination. The disadvantages are inconvenience, cost, and the ability of the airline to open or refuse to carry unaccompanied packages. This may be lessened by the veterinarian requesting "favored shipping" status with the airline; this is required by some airlines for any packages to be shipped from your address. The advantages of using a shipping service are convenience and inability of that service to open a package once it has been checked and sealed. The disadvantage is the necessity of abiding by their shipping schedule. It is up to the individual veterinarian if he or she chooses to take on the responsibility of shipping or asks each client to take on that responsibility him or herself.

Post-care and complications

The semen of some male dogs does not tolerate chilling and shipment, presumably because of a subclinical disease or because of the semen's chemical incompatibility with the extender. It is strongly recommended that dogs intended for use as a stud with chilled semen have semen collected, extended, stored in the refrigerator overnight, and evaluated the next day, to best mimic chilling and shipment. This should be done well before any bitches of interest are in heat. If semen quality is not maintained, problems with the particular male dog can be addressed or another suitable male can be identified to perform the breeding if necessary.

The most common problem with chilled semen is difficulty during shipment. Follow packaging instructions to the letter will minimize logistical problems with the shipper. The breeder is encouraged to trace all package tracking numbers and to have a back-up plan in place should the sample be damaged or lost.

Supplemental reading

England GCW, Millar KM. 2008. The ethics and role of AI with fresh and frozen semen in dogs. *Reprod Dom Anim* 43 (Suppl. 2):165–171.

Lopes G, Simoes A, Ferreira P, et al. 2008. Differences in preservation of canine chilled semen using different transport containers. *Anim Reprod Sci* doi:10.1016/j.anireporsci.2008.04.003.

Pena FJ, Nunez-Martinez I, Moran JM. 2006. Semen technologies in dog breeding: An update. *Reprod Dom Anim* 41 (Suppl. 2):21–29.

Rijsselaere T, Van Soom A, Maes D, et al. 2004. Effect of blood admixture on in vitro survival of chilled and frozen-thawed canine spermatozoa. *Theriogenology* 61:1589–1602.

Root Kustritz MV. 2007. The value of canine semen evaluation for practitioners. *Theriogenology* 68:329–337.

13

What is the technique for preparation of frozen semen in dogs?

Anatomy

Semen is collected as for any other procedure (see Chapter 6).

Pre-procedure considerations

All male dogs used for breeding should be certified free of heritable conditions and infectious disease. All dogs should be tested and proven negative for canine brucellosis before semen is collected for chilling and shipment (see Chapter 44). Specific tests for hereditary conditions vary by breed. Most breeds require Orthopedic Foundation for Animals or University of Pennsylvania Hip Improvement Program certification for hip dysplasia and Canine Eye Registry Foundation evaluation for ocular defects. Breeders should be referred to their national breed club web site or printed materials to see which tests are recommended by the health organization unique to their breed.

If the pups are to be registered with the American Kennel Club (AKC), the male must have a DNA profile on record with the AKC. This rule has been in effect since October 1, 1998. Frozen semen collected and stored prior to October 1, 1998 in accordance with AKC regulations and procedures is exempt from this requirement. The DNA sample is collected using a cheek swab and may be done by the dog's owner. Breeders are referred to the AKC web site for complete information about sample collection, sample submission, and registration and payment (see Resources).

For litters sired using frozen semen to be registered with the AKC, specific regulations regarding identification of pellets or straws of frozen semen and their storage must be followed. Veterinarians wishing to freeze semen should contact the AKC so their site can be inspected and listed through the AKC as a certified freezing center.

Viability of canine spermatozoa is decreased after freezing and thawing. This is due to physical destruction of cells as they undergo osmotic change, dehydration, ice crystal formation, and induction of chemical reactions during the freeze–thaw process that mimics capacitation. Insemination directly into the uterus is strongly recommended. Reported conception rates with frozen–thawed semen for vaginal insemination (see Chapter 14) average 45%, for transcervical insemination (see Chapter 15) 70%, and for surgical intrauterine insemination (see Chapter 16) 95%.

Procedure

Frozen semen is made up of ejaculated spermatozoa suspended in an extender. Extender is a fluid medium containing nutrients, buffers, and anti-microbials and, for frozen semen, a cryoprotectant to minimize disruption of the spermatozoa cells as they freeze and thaw. Cryoprotectants most commonly used are egg yolk and glycerol. Egg yolk may also prevent premature capacitation of spermatozoa. Milk-based extenders have proven successful as alternatives to egg-yolk-based extenders in countries concerned about transmission of avian influenza. Recipes for homemade extenders exist. Most veterinarians use commercial extenders and a commercial system that maintains the semen in long-term storage off-site. Examples of companies for which veterinarians freeze semen include Synbiotics (www.synbiotics.org), CLONE (www.cloneusa.com/chilled.html), and International Canine Semen Bank (ICSB; www.ik9sb.com). ICSB also provides a kit that permits veterinarians to collect semen, extend it, and ship it to ICSB for freezing and storage (see Resources).

Collect semen in a routine manner and perform a complete semen evaluation (see Chapter 7). Only very good- to excellent-quality semen warrants freezing. Samples with high percentages of proximal cytoplasmic droplets may not be suitable for freezing; presence of proximal cytoplasmic droplets has been associated with decreased fertilizing capability and poor survival after freezing and thawing. Semen samples containing 40% or greater blood and blood products are not suitable for cryopreservation; hemoglobin released by hemolysis decreases post-thaw viability. Poor-quality semen may benefit from passage through a filter. With this process, semen is layered over a gradient and centrifuged. Abnormal spermatozoa and blood products are retained and normal spermatozoa gather into a usable pellet. If this technique is used, the centrifugation step described below is skipped.

The following is one example of a technique for semen freezing in dogs. Semen can be frozen in 0.25 or 0.50 ml straws or in pellets, formed when extended semen is deposited in aliquots into depressions on solid dry ice. There are no industry standards for semen freezing in dogs.

Collect semen by manual ejaculation. Determine the total number of spermatozoa in the ejaculate so each straw can be identified as containing "x" number of spermatozoa after dilution in extender and freezing. Centrifuge the semen for about 5 min at the setting used for urine sedimentation (500 to 700 g). Draw up one drop of supernatant and evaluate it for spermatozoa; if many motile spermatozoa are still present in the supernatant, centrifuge the semen for another 5 min. Once centrifugation is complete, pour off the supernatant and resuspend the pellet of semen in room temperature extender at a ratio of 1:2 to 1:4. Chill the sample at refrigerator temperature for 1 h. Label the 0.5 ml polyvinylchloride straws with the animal's AKC registration number and breed, and the date. Add two parts of a second extender, containing glycerol, in four equal aliquots over a 45-min period for a final glycerol concentration of 4%. Fill the straws, leaving an air bubble to prevent expulsive loss of the semen during freezing. Seal the straw and place it at refrigerator temperature for 90 min. Suspend the straws about 5 in above a pool of liquid nitrogen for 5 to 10 min, then slowly lower the straws into liquid nitrogen. Machines that automate this equilibration and lowering process exist; use of such machines is associated with better post-thaw longevity of spermatozoa. At least 5 min later, transfer the straws to a permanent tank containing liquid nitrogen. All samples must be maintained in fluid liquid nitrogen or liquid nitrogen vapor for storage or shipment until they are to be thawed for insemination.

Post-care and complications

The semen of some male dogs does not tolerate freezing, presumably because of a subclinical disease or because of the semen's chemical incompatibility with the extender. It is strongly recommended that dogs intended for use as a stud with frozen semen have semen collected and frozen, and one straw or pellet immediately thawed to assess post-thaw quality. This should be done well before any bitches of interest are in heat.

If semen quality is not maintained, problems with the particular male dog can be addressed and another suitable male can be identified to perform the breeding if necessary. Because some males can be identified as having semen that is difficult to freeze, breeders should be reminded that their unique problem may have a genetic basis and strongly encouraged to be thoughtful about maintaining their dog in the gene pool long-term.

Semen must be maintained in fluid liquid nitrogen or liquid nitrogen vapor at all times. Some shippers refuse to carry canisters containing fluid liquid nitrogen. "Dry shippers" contain liquid nitrogen vapor and will remain charged for 1 to 3 weeks, permitting shipment; either the dry shipper should be replenished with fluid liquid nitrogen on arrival or the straws transferred to a more secure liquid nitrogen container upon arrival.

Supplemental reading

Abe Y, Lee C-S, Sano H, et al. 2008. Artificial insemination with canine spermatozoa frozen in a skim milk/glucose-based extender. *J Reprod Dev* 54:290–294.

Cheng F-P, Wu J-T, Tsai P-S, et al. 2005. Effects of cryo-injury on progesterone receptor(s) of canine spermatozoa and its response to progesterone. *Theriogenology* 64:844–854.

England GCW, Millar KM. 2008. The ethics and role of AI with fresh and frozen semen in dogs. *Reprod Dom Anim* 43 (Suppl. 2):165–171.

Hermansson U, Linde Forsberg C. 2006. Freezing of stored, chilled dog spermatozoa. *Theriogenology* 65:584–593.

Medeiros CMO, Forell F, Oliveira ATD, et al. 2002. Current status of sperm cryopreservation: Why isn't it better? *Theriogenology* 57:327–344.

Pena AI, Barrio M, Becerra JJ, et al. 2008. Infertility in a dog due to proximal cytoplasmic droplets in the ejaculate: Investigation of the significance for sperm functionality in vitro. *Reprod Dom Anim* doi:10.1111/j.1439-0531.2006.00809.x.

Pena A, Linde-Forsberg C. 2000. Effects of spermatozoa concentration and post-thaw dilution rate on survival after thawing of dog spermatozoa. *Theriogenology* 54:703–718.

Rijsselaere T, Van Soom A, Maes D, et al. 2004. Effect of blood admixture on in vitro survival of chilled and frozen-thawed canine spermatozoa. *Theriogenology* 61:1589–1602.

Schafer-Somi S, Kruger S, Knapp E, et al. 2006. Effects of semen extender and semen processing on motility and viability of frozen-thawed dog spermatozoa. *Theriogenology* 66:173–182.

Watson PF. 2000. The causes of reduced fertility with cryopreserved semen. *Anim Reprod Sci* 60–61:481–492.

Witte TS, Schafer-Somi S, Kuchar A, et al. 2009. Effect of hen's egg yolk on capacitation and acrosome reaction of diluted canine spermatozoa. *Anim Reprod Sci* 110:293–205.

14

What is the technique for vaginal insemination?

Anatomy

The vulvar cleft provides entrance to the ventral clitoral fossa and dorsal vestibule. The vestibule is not horizontal in bitches but instead rises at a 45-degree angle relative to the spine. Anything placed in the canine vagina should be introduced dorsally and aimed toward the spine as it is advanced cranially. The vagina of the bitch is very long and the cranial lumen is compressed to nearly one-third in size by the presence of a tissue fold suspended from the dorsal vaginal surface. The cervix is abdominal and the external cervical os hangs from the dorsal surface of the vagina, with a dead-ended fornix lying cranial to it.

Pre-procedure considerations

Vaginal insemination is most commonly used for introduction of fresh or fresh chilled semen. This may be necessary if the bitch and stud do not show compatible breeding behavior, if they are too different in size to permit natural breeding, if the semen is being shipped, or if the owners wish to avoid potential exposure to disease.

All bitches used for breeding should be certified free of heritable conditions and infectious disease. All bitches should be tested and proven negative for canine brucellosis before breeding (see Chapter 44). Specific tests for hereditary conditions vary by breed. Most breeds require Orthopedic Foundation for Animals or University of Pennsylvania Hip Improvement Program certification for hip dysplasia and Canine Eye Registry Foundation evaluation for ocular defects. Breeders should be referred to their national breed club web site or printed materials to see which tests are recommended by the health organization unique to their breed. Progesterone assay and other tests should be used to verify the ovulation date (see Chapter 29). Fresh or chilled semen is usually introduced 2 days after the ovulation date.

Procedure

Pass the gloved finger of the nondominant hand, lubricated with water only, into the vaginal canal. This protects the urethral papilla and prevents accidental introduction of semen into the urinary bladder. Draw semen up into a sterile syringe. Pass a pipette (Fig. 14-1) along the gloved index finger and as far forward into the vagina as possible without causing the bitch discomfort. Sterile pipettes marketed for insemination in dogs are available (see Resources) or one may use

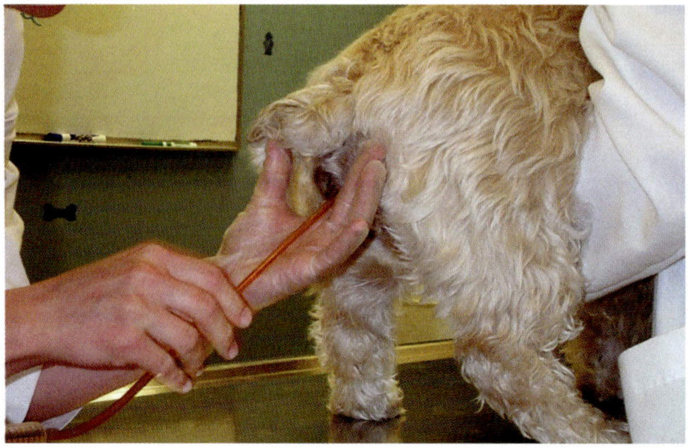

Figure 14-1: Passage of the pipette over the gloved finger into the bitch's vagina (reprinted with permission from Root Kustritz MV. 2006. *The Dog Breeder's Guide to Successful Breeding and Health Management*. St. Louis, MO: Elsevier).

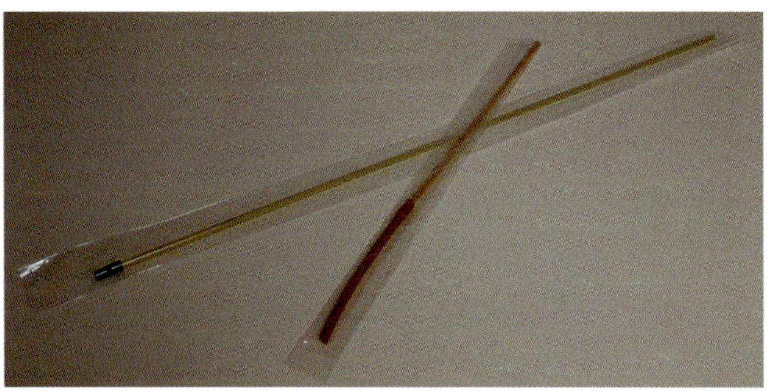

Figure 14-2: Pipettes suitable for intravaginal insemination of bitches (reprinted with permission from Root Kustritz MV. 2006. *The Dog Breeder's Guide to Successful Breeding and Health Management*. St. Louis, MO: Elsevier).

bovine uterine infusion pipettes (Fig. 14-2). Balloon-tipped pipettes are also available; success rate does not vary between plain and balloon-tipped pipettes. Pull the pipette back so you know it is not lodged against a fold of vaginal mucosa and introduce the semen (Fig. 14-3). Semen should flow freely through the pipette. Flush the pipette with a small amount of air. Withdraw the pipette, leaving the gloved finger in place, and use that finger to tickle the ceiling of the vagina. This may stimulate vaginal contractions as would occur during natural breeding and presumably aids in the forward movement of spermatozoa. Withdraw the gloved finger.

The bitch's hindquarters may be elevated for several minutes (Fig. 14-4). No upward pressure should be placed on the bitch's caudal abdomen as she is moved from the examining table or put in the car. She should not be permitted to squat to urinate as she leaves the hospital. All of these instructions may be unnecessary, as it may well be that the spermatozoa that will fertilize the ova are well into the uterus and oviduct within moments of insemination.

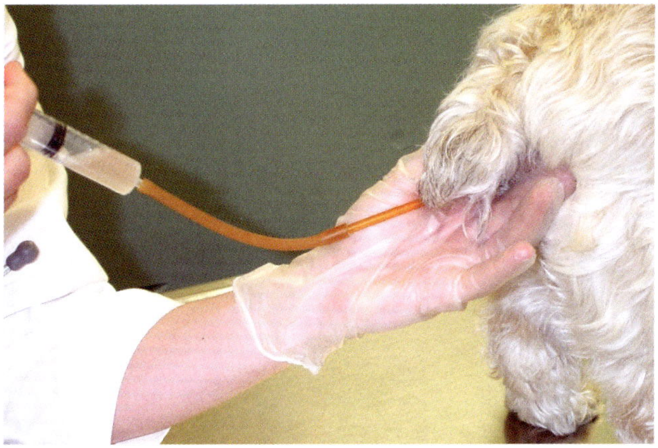

Figure 14-3: Introduction of semen (reprinted with permission from Root Kustritz MV. 2006. *The Dog Breeder's Guide to Successful Breeding and Health Management.* St. Louis, MO: Elsevier).

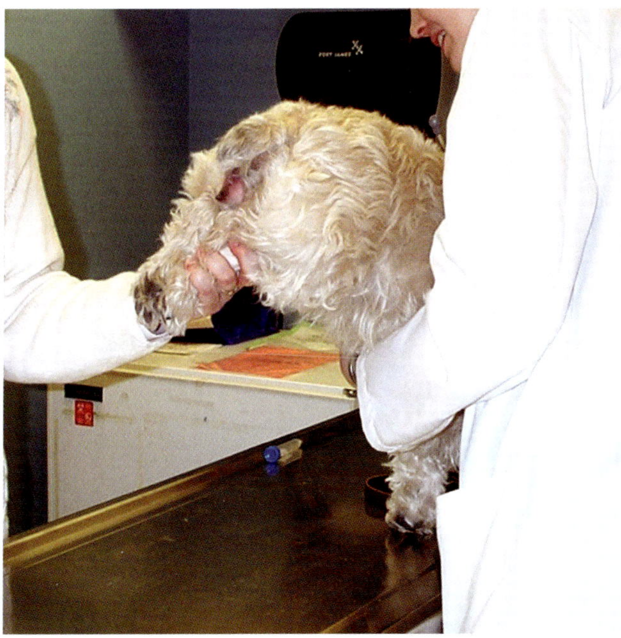

Figure 14-4: "Wheelbarrowing" of the bitch's hindquarters (reprinted with permission from Root Kustritz MV. 2006. *The Dog Breeder's Guide to Successful Breeding and Health Management*. St. Louis, MO: Elsevier).

Post-care and complications

There are occasional reports of vaginal perforation or respiratory distress in bitches restrained for vaginal insemination. These are uncommon. Most bitches are very cooperative. The author has had good success with uncooperative bitches by restraining them on their side for the procedure; however, very anxious bitches often do not conceive.

Supplemental reading

Johnston SD, Root Kustritz MV, Olson PN. 2001. *Canine and Feline Theriogenology*. Philadelphia: WB Saunders Co., pp. 57–65.

Nizanski W. 2006. Intravaginal insemination of bitches with fresh and frozen-thawed semen with addition of prostatic fluid: Use of an infusion pipette and the Osiris catheter. *Theriogenology* 66:470–483.

Thomassen R, Farstad W. 2009. Artificial insemination in canids: A useful took in breeding and conservation. *Theriogenology* 71:190–199.

15

What is the technique for transcervical insemination?

Anatomy

The vulvar cleft provides entrance to the ventral clitoral fossa and dorsal vestibule. The vestibule is not horizontal in bitches but instead rises at a 45-degree angle relative to the spine. Anything placed in the canine vagina should be introduced dorsally and aimed toward the spine as it is advanced cranially. The vagina of the bitch is very long and the cranial lumen is compressed to nearly one-third in size by the presence of a tissue fold suspended from the dorsal vaginal surface. The cervix is abdominal and the external cervical os hangs from the dorsal surface of the vagina, with a dead-ended fornix lying cranial to it.

Pre-procedure considerations

Transcervical insemination most commonly is used for introduction of fresh chilled or frozen semen. The short viability of frozen semen after thawing necessitates it being placed into the uterus and this technique is less invasive than is surgical insemination. Transcervical insemination may also be desirable if quality of fresh semen is poor. Transcervical insemination can be performed multiple times in one heat cycle and can be performed with vaginal or surgical insemination during a given heat cycle.

All bitches used for breeding should be certified free of heritable conditions and infectious disease. All bitches should be tested and proven negative for canine brucellosis before breeding (see Chapter 44). Specific tests for hereditary conditions vary by breed. Most breeds require Orthopedic Foundation for Animals or University of Pennsylvania Hip Improvement Program certification for hip dysplasia and Canine Eye Registry Foundation evaluation for ocular defects. Breeders should be referred to their national breed club web site or printed materials to see which tests are recommended by the health organization unique to their breed. Progesterone assay and other tests should be used to verify the ovulation date (see Chapter 29). Fresh and chilled semen is usually introduced 2 days after the ovulation date. Frozen semen is usually introduced 3 to 4 days after ovulation.

Procedure

A rigid, narrow-diameter endoscope, as is used for cystocopy, is required (see Chapter 3, Fig. 3-2). Introduce the endoscope dorsally at the vulvar cleft and move it gently forward at a 45-degree angle, then angle it horizontally. The vaginal mucosa should be blanched and crenated (sharply wrinkled; see Chapter 29, Fig. 29-1). Identify the caudal end of the post-cervical dorsal median fold as a compression from the ceiling of the vagina narrowing the vaginal vault. Advance the scope to one side or the other. When the dead-ended fornix is seen, the cervix should be visible dorsally, appearing as a rosette of folds from which issues the serosanguinous or straw-colored discharge of estrus.

Pass a long, sterile, polypropylene urinary catheter through the biopsy channel of the scope, such that its tip comes out at the tip of the endoscope. Rest the tip of the endoscope against the external cervical os and advance the catheter into the cervical lumen. This is most easily achieved by spinning the catheter between the fingers. Several attempts may be necessary. Once the catheter is in place, frozen semen should be thawed. Semen should be drawn into a sterile syringe and introduced through the catheter. The semen should flow easily and be visible flowing through the catheter into the uterus with minimal backflow. Withdraw the catheter and the endoscope.

Post-care and complications

There are occasional reports of vaginal perforation during passage of the long rigid endoscope. This occurs very rarely during estrus, when the vaginal epithelium is thick and cornified. Most bitches are very amenable to the procedure; however, if a bitch is jumpy or resists restraint, the long, rigid endoscope should not be passed.

The pioneer of this technique, Dr. Marian Wilson, reports success in dogs ranging from toy to giant breeds. There are some bitches whose anatomy is such that passage of the catheter by this route is not possible. In those bitches, vaginal or surgical insemination is recommended.

Supplemental reading

Thomassen R, Farstad W. 2009. Artificial insemination in canids: A useful tool in breeding and conservation. *Theriogenology* 71:190–199.

Wilson MS. 2001. Transcervical insemination techniques in the bitch. *Vet Clin North Am* 31:291–304.

16

What is the technique for surgical insemination?

Anatomy

The uterus of the bitch has a small body and long, torturous horns. Semen can be deposited anywhere within the lumen of either horn.

Pre-procedure considerations

Surgical insemination is used to introduce semen directly into the uterus, bypassing the vagina and cervix. It is most commonly used with frozen-thawed semen but it can also be used for breeding bitches with a history of subfertility or when using semen of questionable quality. Although canine oocytes may remain viable for more than 200 h after ovulation, cervical closure occurs 6 to 8 days after the luteinizing hormone peak, requiring introduction of semen directly into the uterus.

All bitches used for breeding should be certified free of heritable conditions and infectious disease. All bitches should be tested and proven negative for canine brucellosis before breeding (see Chapter 44). Specific tests for hereditary conditions vary by breed. Most breeds require Orthopedic Foundation for Animals or University of Pennsylvania Hip Improvement Program certification for hip dysplasia and Canine Eye Registry Foundation evaluation for ocular defects. Breeders should be referred to their national breed club web site or printed materials to see which tests are recommended by the health organization unique to their breed. Progesterone assay and other tests should be used to verify the ovulation date (see Chapter 29). Frozen semen is usually introduced into the female 3 to 4 days after ovulation. Preparation and diagnostic testing should be completed as for any procedure requiring general anesthesia.

Procedure

Place the animal under general anesthesia and shave and prepare the abdomen for sterile surgery. Make a ventral midline incision and exteriorize the uterus. Pass a 22-gauge needle or catheter into the lumen of the uterine body (Fig. 16-1).

If frozen semen is being used, semen should be thawed using the instructions provided by the person who froze that semen. Semen should be thawed as close to the moment of insemination as possible to minimize the time the thawed semen is held at room temperature.

Draw the semen up into a sterile syringe. Introduce the semen through the catheter; it should flow freely into the uterine lumen. Even small volumes distend the uterine body and both horns.

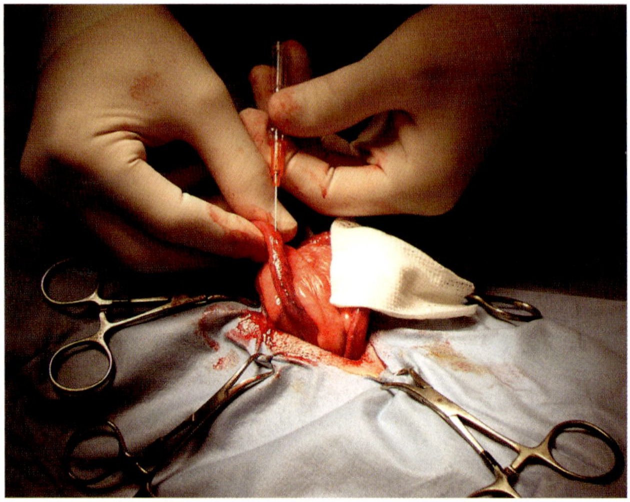

Figure 16-1: Insertion of a catheter into the lumen of one uterine horn (photo courtesy of Inver Grove Heights Animal Hospital).

Muscle fasciculations of the uterus may be evident after the semen is introduced. Withdraw the needle or catheter and blot the hole with gauze until hemostasis is achieved. Replace the uterus in the abdomen and close in a routine manner.

Post-care and complications

Incision care is as for any abdominal incision.

Supplemental reading

Boland P. 2004. Surgical insemination and other ways. Proceedings, European Veterinary Society for Small Animal Reproduction, Barcelona, Spain.

Johnston SD, Root Kustritz MV, Olson PNS. 2001. *Canine and Feline Theriogenology*. Philadelphia: WB Saunders Co., pp. 62–63.

Thomassen R, Farstad W. 2009. Artificial insemination in canids: A useful tool in breeding and conservation. *Theriogenology* 71:190–199.

Verstegen JP, Silva LD, Onclin K. 2001. Determination of the role of cervical closure in fertility regulation after mating or artificial insemination in Beagle bitches. *J Reprod Fertil* 57:31–34.

17

What is the technique for anesthesia for Cesarean section?

Anatomy

Anatomic changes of pregnancy are reflected in physiologic changes, as described below.

Pre-procedure considerations

Physiologic changes of pregnancy that impact anesthesia include an increase in plasma volume and decrease in hematocrit, increased oxygen consumption and carbon dioxide production, increased cardiac output, decreased blood pressure due to vasodilation and hypotension due to mechanical compression from the gravid uterus, and slowed gastrointestinal (GI) motility. The clinical significance of these changes are the need to address hydration to maintain normal blood pressure during surgery, and the need for a well-fitted, cuffed endotracheal tube to counter aspiration of GI contents.

Once a decision has been made to perform Cesarean section (C-section), time is of the essence. An intravenous catheter should be placed and fluid therapy instituted. The surgical site should be clipped and prepped as much as possible before any anesthetic agents are introduced. Equipment for neonatal resuscitation should be assembled including oxygen and small masks or endotracheal tubes, hemostats to clamp off the umbilical cords, suction apparatuses to remove fluid from the oral and nasal cavities, warm towels, and appropriate medications (see below).

Procedure

Considerations of anesthesia

Maternal stability under anesthesia should be monitored by assessment of heart rate and rhythm, blood pressure, and oxygenation status (pulse oximetry), if possible. Maintaining maternal blood pressure is an important component in maintaining placental blood flow and supporting the pups until they are delivered. Adequately oxygenate the bitch at all times.

Choosing appropriate anesthetic agents

Remember that most anesthetic agents exert effect because they are soluble in fat and move readily into tissue. That means that most will readily cross the placenta and that anything given to the dam is also given to the pups. Choose drugs that cause minimal cardiovascular depression or that are easily reversed. Short-acting drugs will have less of an effect on the fetuses as they are

Table 17-1. Drugs used for anesthesia of bitches for Cesarean section.

Drug Class	Representative Drugs	Pros	Cons
Opioids	Buprenorphine, butorphanol, hydromorphone, morphine, oxymorphone	Reversible with naloxone	May cause respiratory depression in dam and pups, buprenorphine difficult to reverse.
Benzodiazepines	Diazepam, midazolam	Reversible with flumazenil	May cause profound sedation in pups.
Alpha-2-adrenergic agonists	Medetomidine, detomidine, xylazine	—	May cause profound sedation in pups, may cause bradycardia and arrhythmias in dam, may decrease uterine blood flow leading to fetal hypoxemia.
Phenothiazine	Acepromazine	—	May cause profound long-lasting sedation in pups, not reversible, may cause hypotension in dam.
Dissociatives	Ketamine, tiletamine (Telazol)	—	Not reversible, may cause uterine vasoconstriction leading to fetal hypoxemia.
Thiobarburitate	Thiopental	Rapidly distributed and can be metabolized by fetal liver	Not reversible.
—	Propofol	Very short-acting	Not reversible, may cause transient apnea in the dam.
—	Etomidate	Minimal cardiovascular effects on mother and pups	Not reversible, relatively expensive.
Anticholinergics	Atropine, glycopyrrolate	Atropine can increase fetal heart rate—this may be beneficial if countering drugs given to dam	Atropine can increase fetal heart rate—fetal tachycardia not always beneficial, glycopyrrolate less likely to cross the placenta and show fetal effects.
Local anesthetics	Bupivicaine, lidocaine	Epidural and/or line block provides pain control and minimizes need for systemic drugs	Vasodilation in dam may precipitate hypotension, if dam agitated may be difficult to maintain sterile field.
Inhalants	Isoflurane, sevoflurane, halothane	Can minimize fetal depression by altering dose quickly	All cross the placenta rapidly.
—	Nitrous Oxide	Use permits use of lower doses of more potent inhalants	Hypoxia in dam and pups.

Table 17-2. Sample anesthetic protocols for canine Cesarean section.

Protocol 1—If the mother is calm and quiet, and can be catheterized without sedation
* Induce anesthesia with propofol to effect.
* Intubate and maintain a light plane of anesthesia with sevoflurane or isoflurane in oxygen.
* Line block or epidural for pain management.
* After neonates are delivered, add hydromorphone or oxymorphone IV for additional maternal pain control.

Protocol 2
* Sedate with fentanyl and diazepam.
* Epidural containing bupivacaine and morphine.
* Provide oxygen via facemask.

Protocol 3
* Mask induction with sevoflurane or isoflurane in oxygen.
* Intubate and maintain a light plane of anesthesia with sevoflurane or isoflurane in oxygen.
* After neonates are delivered, add hydromorphone or oxymorphone IV for maternal pain control.

more rapidly cleared; longer acting drugs for the sake of the dam can be added after the pups are delivered.

Pregnant animals have an increased pain threshold due to release of endogenous endorphins. When considering anesthetic agents, generally this translates to a reduction in inhalant anesthesia requirements and reduced need for opioids to achieve desired analgesia. If epidural anesthesia is desired, venous engorgement in the epidural space and increased epidural pressure are reflected in the need for lower doses of local anesthetics to achieve desired effect. The pros and cons of various drugs are described in Table 17-1. Sample protocols are outlined in Table 17-2.

Post-care and complications

The bitch is recovered from anesthesia and extubated in a routine fashion.

Puppy mortality is greater with C-section than with free-whelping, and within C-sections, is greater for emergency than for elective surgeries. Specific steps that can be taken to enhance success in puppy resuscitation include (1) appropriate selection of anesthetic agents, as described above, and (2) implementation of a sequence of steps to be taken by all personnel resuscitating puppies (Table 17-3). Although it has been demonstrated that puppy vitality is lower at birth in pups born through C-section than in pups born vaginally with no complications or born vaginally in dystocia, appropriate resuscitation should lead to normalization within 5 min of life.

Supplemental reading

Funkquist PME, Gorel CN, Lofgren AJ, et al. 1997. Use of propofol isoflurane as an anesthetic regimen for Cesarean section in dogs. *J Am Vet Med Assoc* 211:313–317.

Silva LCG, de Fatima Lucio C, Veiga GAL, et al. 2008. Neonatal clinical evaluation, blood gas and radiographic assessment in normal birth, vaginal dystocia or cesarean section in dogs. Proceedings, International Syposium on Canine and Feline Reproduction, Vienna, Austria.

Luna SP, Cassu RN, Castro GB, et al. 2004. Effects of four anesthetic protocols on the neurological and cardiorespiratory variables of puppies born by caesarean section. *Vet Rec* 154:387–389.

Moon PF, Erb HN, Ludders JW, et al. 1998. Perioperative management and mortality rates of dogs undergoing cesarean section in the United States and Canada. *J Am Vet Med Assoc* 213:365–369.

Table 17-3. Puppy resuscitation.

1. Keep the puppy warm. Dry by vigorously rubbing the pup with a warm, dry towel. Clear the mouth and nose of fluid by suction.
2. If opioids were given to the dam, reverse with naloxone (one drop from a 25-gauge needle, sublingually). If benzodiazepines were given to the dam, reverse with flumazenil.
3. If the neonate is breathing at least 10 bpm and moving or crying, finish drying the pup, place it in a warm area (warm air circulating pad or radiant heat), and clamp or tie off the umbilicus. If the pup is NOT breathing at least 10 bpm and moving or crying, provide oxygen for 30 to 40 s and continue tactile stimulation by rubbing.
4. If the pup does not begin to breathe spontaneously, attempt to stimulate respiration with placement and rotation of a 25-gauge needle into the nasal philtrum (GV 26 or Jen Chung acupuncture site). If that does not work, consider administration of doxapram sublingually.
5. If there is no discernible heart rate, begin gentle chest compressions (1 to 2 per second).
6. If oxygenation and chest compressions do not elicit spontaneous breathing and heart beat, consider trying placement of an intraosseous catheter (see Chapter 90) and administration of intravenous 10% dextrose solution (2 to 4 ml/kg as a slow bolus) or sodium bicarbonate (1 ml/kg of the 1 IU/ml concentration).
7. If there is no response by 30 min, the pup is declared dead.

Moon PF, Erb HN, Ludders JW, et al. 2000. Perioperative risk factors for puppies delivered by cesarean section in the United States and Canada. *J Am Anim Hosp Assoc* 36:359–368.

Moon-Massat PF, Erb HN. 2002. Perioperative factors associated with puppy vigor after delivery by cesarean section. *J Am Anim Hosp Assoc* 38:90–96.

Pascoe PJ, Moon PF. 2001. Periparturient and neonatal anesthesia. *Vet Clin North Am* 31:315–341.

Ryan SD, Wagner AE. 2006. Cesarean section in dogs: Physiology and perioperative considerations. *Compendium* 28:34–42.

Traas A. 2008. Resuscitation of canine and feline neonates. *Theriogenology* 70:343–348.

18

What is the technique for Cesarean section?

Anatomy

The ovaries lie caudal to the kidneys and each is encased in an ovarian bursa. The primary blood supply is the ovarian artery. The body of the uterus lies within the pelvis; the uterine horns lie in the abdomen. The uterus and ovaries are suspended and attached to the body wall by the broad ligament. The primary uterine blood supply is the bilateral uterine arteries, which arise from the vaginal arteries caudally and lie along the uterine body with small branches supplying the uterine horns over their length.

Pre-procedure considerations

Cesarean section (C-section) is indicated if obstructive dystocia is present, if the bitch is exhibiting primary or secondary uterine inertia, if medical therapy for dystocia has been unproductive, or if fetal heart rate is less than 170 bpm (see Chapter 39). Concurrent ovariohysterectomy (OHE) should be offered to the owner, especially if uterine tissue does not appear viable. The primary advantage is that the bitch will not have to undergo another anesthetic episode for OHE in the future. The disadvantages are an increased risk of hemorrhage and hypovolemia, and possible increase in length of hospital stay. Milk production will not be altered by concurrent OHE.

The bitch should be prepared for surgery to the greatest extent possible before any anesthetic medications are administered. Details of anesthesia are presented in Chapter 17.

Procedure

The bitch is shaved and prepped for sterile surgery. Place the bitch in dorsal recumbency; ensure that the head is not lower than the abdomen, as pressure on the diaphragm will compromise ventilation. Both ventral midline and flank approaches have been described; the former is more common. Make an incision from just cranial to the pubis to the umbilicus. The abdominal musculature is stretched thin; use caution when incising the linea alba.

Exteriorize the gravid uterus and pack it off with saline-moistened laparatomy sponges (Fig. 18-1). Make an incision in a relatively avascular area of the uterus in one horn near the uterine body. All pups can be milked from both horns to this single incision readily (Fig. 18-2). If there is a large number of pups, an incision may be made in each uterine horn. If a pup is stuck in the pelvic canal, an incision should be made in the uterine body.

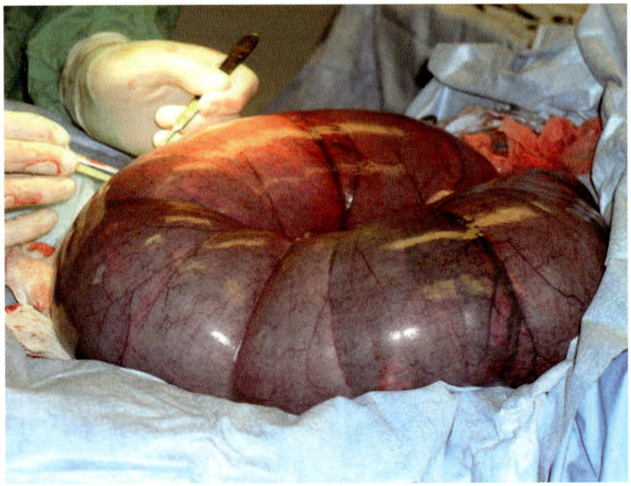

Figure 18-1: Exteriorized gravid canine uterus (photo courtesy of Inver Grove Heights Animal Hospital).

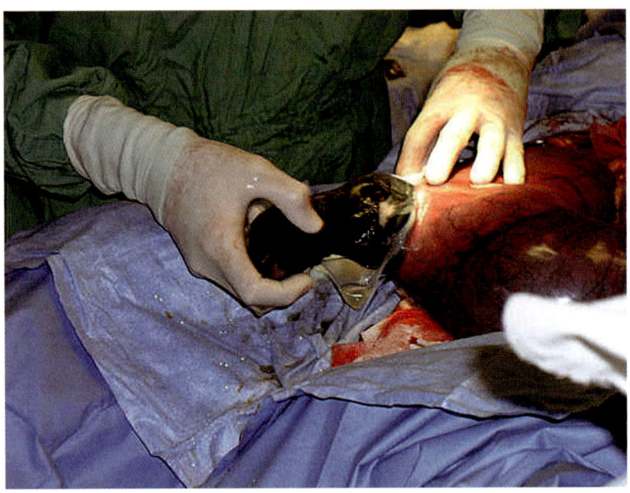

Figure 18-2: Removing pup through uterine incision (photo courtesy of Inver Grove Heights Animal Hospital).

The pup is removed, still encased within its amniotic sac, and handed off to waiting personnel for resuscitation (see Chapter 17). The placenta and other fetal tissues should be removed with the pup if they come readily. If the placenta is still tightly adhered to the endometrium, it may minimize bleeding if it is left in place and allowed to detach and pass with postpartum lochia.

When all pups have been removed, palpate the entire uterus and into the pelvic canal to make sure no pups have been missed. If the bitch is not to be spayed, close the uterus with an absorbable suture in an inverting pattern, in one or two layers. If the bitch is to be spayed, the uterus is roughly closed in a continuous pattern to minimize leaking of uterine contents into the abdomen during manipulation, and routine OHE is performed. Flush and inspect the abdomen.

Close the abdominal layers routinely. Nonabsorbable suture may be used to ensure closure and healing of the thinned linea alba.

En-bloc OHE has been described as a means of performing concurrent OHE and C-section in bitches. The ovarian arteries and uterine arteries and body are clamped and the gravid uterus removed. The uterus is handed off to an assistant who quickly opens the uterus and removes all the pups for resuscitation. The surgeon returns to the bitch to ligate the ovarian and uterine vessels and complete the OHE. The advantage of this technique is minimal transfer of anesthetic agents to the pups and decreased possibility of abdominal contamination with uterine contents. The disadvantage is the possibility of fetal stress secondary to loss of uterine and placental blood flow. There are few studies documenting success rates with this technique compared to standard C-section.

Post-care and complications

Little is known about pharmacokinetics of pain medications in lactating bitches. Nonsteroidal anti-inflammatory drugs do not reach significant concentrations in milk but are contraindicated in humans breast-feeding premature infants as the Cyclooxygenase-2 enzymes inhibited by these drugs are necessary for normal neonatal renal development. Opioids enter milk but total dose ingested by any given pup is low so this may be the preferred drug class for this indication.

Necessity of C-section to relieve dystocia at one parturition is not indicative of a requirement for C-sections at future whelpings. For most bitches that require more than one C-section it is because there is a common underlying cause for dystocia in both cases or because they are of a breed that does not readily free-whelp.

Supplemental reading

Ryan SD, Wagner AE. 2006. Cesarean section in dogs: Physiology and perioperative considerations. *Compendium* 28:34–42.

Traas AM. 2008. Surgical management of canine and feline dystocia. *Theriogenology* 70:337–342.

Section II

Canine reproductive physiology

19

What is the normal age for puberty onset in bitches and dogs?

General information

Puberty is defined in bitches as first proestral bleeding or heat. Normal breeding behavior is expected but rarely tested as few bitches are purposefully bred on their first heat. The average age of first estrus in bitches is 6 to 10 months with a wide normal range varying from 4 months to 2 years. Small breed bitches cycle earlier in life than do large and giant breed bitches. All bitches are considered abnormal if they have not exhibited obvious signs of estrus by 24 months of age (see Chapter 52).

Puberty is defined in male dogs as normal semen quality and exhibition of normal breeding behavior. The average age of puberty in dogs is 10 months with a wide normal range varying from 4 to 5 months to 2 years. Small breed dogs go through puberty earlier in life than do large or giant breed dogs. Beagles have been demonstrated to have normal spermatogenesis by 8 to 9 months of age. All intact male dogs should have normal semen quality by 24 months of age (see Chapter 56). Normal breeding behavior may or may not be present by 24 months of age; many nonmedical factors affect breeding behavior.

Clinical implications

Attention should be paid to the age of female and male dogs that are to undergo gonadectomy prior to puberty onset, with ovariohysterectomy or castration scheduled at about 4 to 6 months for toy and small breeds. Intact females and males should not be housed together unsupervised after 6 to 10 months of age if breeding is not desired. Wild dog crosses and one domestic breed, the Basenji, cycle seasonally. Basenjis cycle in the fall in the northern hemisphere and may come into heat slightly young if it is the right time of year.

Supplemental reading

Goedken MJ, Kerlin RL, Morton D. 2008. Spontaneous and age-related testicular findings in Beagle dogs. *Toxicol Pathol* 36:465–471.

Johnston SD, Root Kustritz MV, Olson PN. 2001. *Canine and Feline Theriogenology*. Philadelphia: WB Saunders Co., pp. 18, 276–277.

20

What are the normal parameters for the estrous (heat) cycle in bitches?

General information

The heat or estrous cycle consists of four stages. These are proestrus, estrus, diestrus, and anestrus. Bitches are monoestrus and are nonseasonal; they do not cycle continuously and do not cycle in response to season. The only exception to this is the Basenji, which cycles in autumn in the northern hemisphere.

Proestrus is the stage where the bitch first exhibits physical and behavioral changes of heat. Physical changes are turgid swelling of the vulva and passage of serosanguinous vulvar discharge (Fig. 20-1). This discharge arises in the uterus and should not be frankly hemorrhagic or have a foul odor. Male dogs will be attracted to the proestrous bitch but she will not allow the male to mount her. Proestrus lasts an average of 9 days but may range from 0 to 17 days.

Estrus, or standing heat, is the stage where the bitch allows the male to mount and breed her. The classic signs include softening of the vulva and change in color of vulvar discharge from serosanguinous to straw-colored at estrus onset (Fig. 20-2). However, many normal, fertile bitches do not show these physical changes. Some estrous bitches will mount other females. The male will investigate the bitch's hindquarters and she will allow him to mount and breed her. Estrus lasts an average of 9 days but may range from 3 to 21 days.

Diestrus is the stage during which successfully bred bitches are pregnant. Most species have a mechanism in place for maternal recognition of pregnancy such that nonpregnant females cut this stage of the cycle short and return to proestrus and estrus repeatedly unless that cycle is broken by pregnancy, disease, or season. Bitches do not recognize pregnancy but instead all bitches that go through estrus continue through a prolonged diestrus whether they were bred or not and whether they are pregnant or not. A milky, odorless discharge may be seen very early in diestrus but, in general, vulvar discharge should cease during this stage. Because all the hormonal changes of pregnancy take place, even nonpregnant bitches may show mammary development and lactation late in diestrus. Male dogs are no longer interested. Some bitches exhibit mild lethargy during this stage. Diestrus lasts about 60 days.

Anestrus is the stage of reproductive quiescence. Underlying hormonal changes during this stage are characterized by increasing frequency and amplitude of gonadotropin secretion, mediated by dopamine. There are no specific physical or behavioral changes. Anestrus lasts about 4.5 months in the average bitch but may vary widely. The average bitch begins a new proestrus every 7 months.

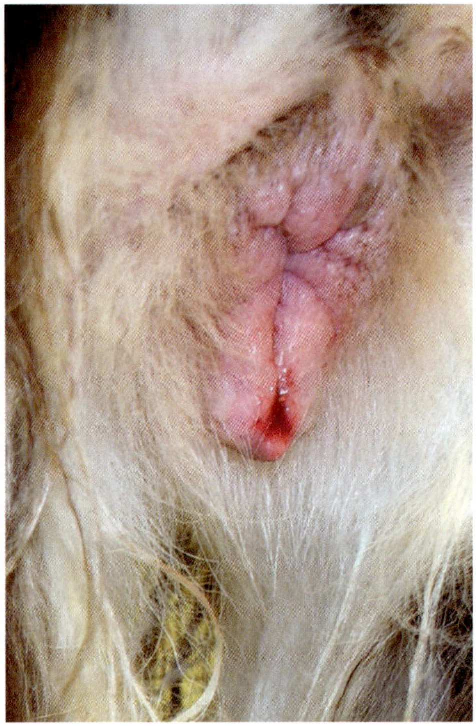

Figure 20-1: Swollen vulva and serosanguinous vulvar discharge characteristic of proestrus (reprinted with permission from Root Kustritz MV. 2006. *The Dog Breeder's Guide to Successful Breeding and Health Management.* St. Louis, MO: Elsevier).

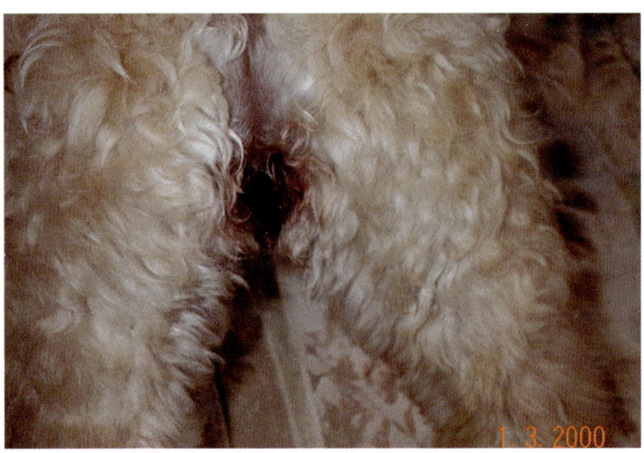

Figure 20-2: Softer vulva and tan-colored discharge characteristic of estrus (reprinted with permission from Root Kustritz MV. 2006. *The Dog Breeder's Guide to Successful Breeding and Health Management.* St. Louis, MO: Elsevier).

Clinical implications

There are no tests available that permit us to predict when a bitch is going to enter proestrus. Experienced male dogs become more interested in the urine of female dogs just before they exhibit signs of proestrus, investigating where they urinate and urinating over it as if it hides evidence of the proestrous bitch from other males.

Many breeders expect to breed their bitch on days 9, 11, and 13 of their cycle, counting the first day of proestrus as day 1. This is based on the fact that the average bitch is in proestrus for 9 days and ovulates on the second day of estrus. Bitches bred on days 9, 11, and 13 are then bred prior to ovulation, on ovulation day, and again 2 days after ovulation. However, many normal bitches are not average bitches. Uncooperative bitches should be evaluated to determine if they are at the appropriate stage of the cycle (see Chapter 29) or if they have a vaginal anomaly or some other physical impediment to breeding (see Chapter 52).

Because all bitches go through a complete diestrus, with hormonal and physical changes similar to those in pregnant bitches, pregnancy diagnostic techniques commonly used in other species do not work well in bitches (see Chapter 5). The prolonged diestrus of bitches predisposes them to pyometra (see Chapter 47) and false pregnancy.

False pregnancy is a misnomer; the condition is more correctly termed pseudocyesis or false whelping. Clinical signs appear concurrent with decrease in serum progesterone concentration at the end of diestrus and include mammary development, lactation, mothering of inanimate objects, and possible aggression. Most clinical signs resolve without therapy. Wrapping the mammary area with an elastic bandage minimizes licking at the glands and increases mammary pressure, which inhibits prolactin production. This, coupled with cutting food and water in half for a day, may hasten resolution. Prolactin-inhibiting drugs can be used to inhibit lactation; the only such drug approved for use in dogs is cabergoline (5 μg/kg once daily for 5 to 10 days). If a sedative is needed to quiet an aggressive bitch, a benzodiazepine such as diazepam should be used in preference to a phenothiazine such as acepromazine, as the latter may stimulate prolactin release.

Anestrus averages 4.5 months, such that the average bitch cycles, or begins a new proestrus, every 7 months. Some breeds, including the German Shepherd dog and Rottweiler, cycle as frequently as every 4.5 to 5 months. Some lines of dogs cycle less frequently. Bitches are not considered abnormal unless they cycle less frequently than every 12 months (see Chapter 52).

Supplemental reading

Gobello C, De la Sota RL, Goya RG. 2001. A review of canine pseudocyesis. *Reprod Dom Anim* 36:283–288.

Johnston SD, Root Kustritz MV, Olson PN. 2001. *Canine and Feline Theriogenology*. Philadelphia: WB Saunders Co., pp. 16–31.

Okkens AC, Kooistra HS. 2006. Anoestrus in the dog: A fascinating story. *Reprod Dom Anim* 41:291–296.

21

What are the normal parameters for semen quality in dogs?

General information

Semen collection and evaluation are described in Chapters 6 and 7, respectively. Semen evaluation in dogs includes assessment of color, volume, percentage of progressively motile spermatozoa, concentration and total number of spermatozoa in the ejaculate, and percentage of morphologically normal spermatozoa. Other parameters that may be evaluated include pH, concentration of alkaline phosphatase in seminal fluid, cytology, and microbiological culture (Table 21-1).

Clinical implications

Bright or dark red coloration is due to fresh blood contamination, most commonly due to prostate disease (see Chapters 53 and 54) or penile trauma (Fig. 21-1). Breaking of multiple small vessels on the surface of the penis may be seen at a young dog's first semen collection. Brown discoloration is usually indicative of old blood, associated with prostate disease. Yellow color is urine contamination. Green discoloration is indicative of prostate infection (see Chapter 54).

Volume varies with collector and is not a reflection of the dog's capabilities. Volume must be noted, however, as it is concentration multiplied by volume that yields total number of spermatozoa in the ejaculate.

Morphologic defects can be classified as primary (occurring during spermatogenesis) or secondary (occurring during storage in the epididymis or as an artifact of sample preparation). Examples of primary morphologic defects are double heads, double midpieces, double tails, bent midpieces, and proximal cytoplasmic droplets. Examples of secondary defects are bent tails, detached heads, and distal cytoplasmic droplets. Effect of a preponderance of any given defect on fertility is not well defined in dogs. If most defects are primary, this suggests that a problem lies within the testis and prognosis for return to normal semen quality is worse than if most defects are secondary.

The wide range of normal for the total number of spermatozoa in the ejaculate is due to the wide range of testicular size in domestic dogs. The minimum number of normal spermatozoa needed to impregnate bitches is about 250 million. A dog with 500 million spermatozoa in his ejaculate would probably be fertile even if only 50% were morphologically normal (500 million $\times$ 0.5 = 250 million). However, if that same male only had 10% morphologically normal

Table 21-1. Normal semen quality parameters in dogs.

Parameter	Normal
Color	Milky or opalescent
Volume	1 to 30 ml
Percentage of progressively motile spermatozoa	70% or greater
Total number of spermatozoa in the ejaculate	300 million to 2 billion
Percentage of morphologically normal spermatozoa	80% or greater
pH	6.3 to 6.7
Alkaline phosphatase of seminal fluid	>5000 IU/l
Cytology	Noninflammatory; low numbers of healthy polymorphonuclear cells may be present in normal semen.
Microbiological culture	Fewer than 10,000 bacteria/ml is nonsignificant bacterial growth.

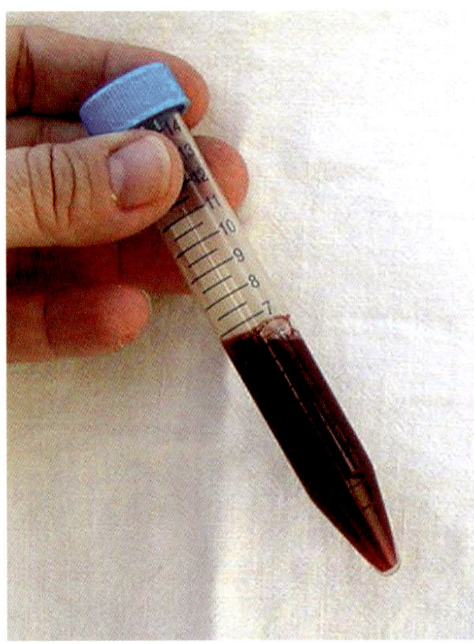

Figure 21-1: Red discoloration of semen associated with prostate disease.

spermatozoa, he would have to breed a given bitch five times over her fertile period to be likely to achieve pregnancy.

Alkaline phosphatase (ALP) in seminal fluid arises from the epididymes and testes. Samples of seminal fluid containing no spermatozoa should be assessed for ALP concentration. Low ALP concentration value suggests that no fluid was retrieved from the reproductive tract, implicating that apprehension or pain is preventing complete ejaculation, or obstruction. On the other hand, normal value (>5000 IU/l) suggests that fluid was retrieved from the reproductive tract and that spermatogenesis is not occurring.

Cytologic changes and growth of bacteria from seminal fluid are not well correlated. If prostatitis or infection elsewhere in the reproductive tract is suspected, microbial culture should be performed even if cytology is noninflammatory.

Supplemental reading

Root Kustritz MV, Johnston SD, Olson PN, et al. 2005. Relationship between inflammatory cytology of canine seminal fluid and significant aerobic, anaerobic or mycoplasma cultures of canine seminal fluid: 95 cases (1987–2000). *Theriogenology* 64:1333–1339.

Root Kustritz MV. 2007. The value of canine semen evaluation for practitioners. *Theriogenology* 68:329–337.

Section III

Canine reproductive management

22

What age is best for ovariohysterectomy of female dogs?

General information

While studies have demonstrated the safety of ovariohysterectomy (OHE), or spay, in dogs as young as 7 weeks of age, there are no prospective long-term studies demonstrating the optimal age for OHE of female dogs. Non-owned dogs (stray or feral animals, humane society animals) should be spayed prior to placement in a new home. Owned animals should be assessed on a case-by-case basis, weighing the pros and cons. There is much in the popular literature on this topic; veterinarians are cautioned to be aware of the number and validity of studies cited to support statements of fact in any publication.

Pros of OHE include a decrease in incidence of mammary neoplasia when aged; decreased incidence of uterine disease, most notably pyometra; and decreased incidence of sexually dimorphic behaviors. The sexually dimorphic behavior most likely to be reduced by OHE in female dogs is inter-female aggression, most commonly displayed during heat.

Mammary neoplasia is the most common tumor type in female dogs, with a reported incidence of 3.4%. Half of the reported cases are malignant, with spread locally and to regional lymph nodes, lungs, and other tissues. Multiple studies over decades have documented a protective effect of OHE; it has been reported that sexually intact female dogs have seven times the risk of developing a mammary tumor when aged compared with spayed dogs and that female dogs spayed prior to their first estrus have 0.5% the risk of developing mammary neoplasia compared with females left intact. Prevention is preferred to treatment. Surgery is the preferred treatment modality; it was reported in one study that 60% of dogs are euthanized at the time of diagnosis.

Pyometra is common in aged female dogs, with a reported incidence of 15.2% in females aged 4 years, and 23 to 24% in females 10 years or older. OHE cures pyometra, with reported mortality rates ranging from 0 to 17%.

Cons of OHE include general risks of surgery, reported in bitches with an incidence of 7 to 27%; urinary incontinence; weight gain; predisposition to cranial cruciate ligament (CCL) injury; and possible predisposition to several uncommon, high morbidity cancers including osteosarcoma, transitional cell carcinoma, and hemangiosarcoma.

Urinary incontinence associated with OHE is urethral sphincter mechanism incompetence, formerly called estrogen responsive urinary incontinence. It was demonstrated statistically in a large retrospective study evaluating effects of gonadectomy that female dogs spayed before 3 months of age were significantly more likely to develop this common, low morbidity disorder than were dogs spayed after 3 months of age.

Obesity is the most common nutritional disorder in dogs and multiple retrospective studies have demonstrated a correlation between spay and weight gain. In cats, it has been demonstrated that gonadectomy is associated with decreased metabolic rate and it may well be that the same is true in dogs. Weight gain is easily controlled by the owner, with appropriate diet and exercise.

CCL injury is more common in women than in men and more commonly occurs during specific stages of the menstrual cycle, suggesting that hormonal factors may play a role. Several studies have demonstrated increased incidence of CCL injury in spayed or neutered dogs, compared with intact dogs; obesity was not indicated as a contributing factor in these studies. Studies have not demonstrated asynchrony in bone growth secondary to gonadectomy, or excessive angulation of the tibial plateau as causative factors.

Transitional cell carcinoma is the most common tumor of the urinary tract in dogs, with a reported incidence of about 1%. Two studies have demonstrated increased incidence in spayed dogs compared with intact female dogs. Similarly, increased incidence of osteosarcoma, a highly malignant tumor with a reported incidence of 0.2%, was reported in spayed populations compared with intact populations in two studies, one of which evaluated the disease in Rottweilers, which have a hereditary predisposition to osteosarcoma. Hemangiosarcoma is another highly malignant, low-incidence tumor reported to have increased incidence in spayed dog populations in two studies. No cause-and-effect relationship has been defined for any of these tumors.

Clinical implications

Veterinarians and clients must take into account incidence, and morbidity and mortality of disorders associated with OHE to determine optimal age of OHE for a given dog. Breed may also play a role. The author strongly recommends OHE after 3 months of age but before onset of first estrus in all female dogs not intended for breeding.

Supplemental reading

Romagnoli S. 2008. Surgical gonadectomy in the bitch and queen: Should it be done and at what age? Proceedings, Southern European Veterinary Conference, Barcelona, Spain.

Root Kustritz MV. 2007. Determining the optimal age for gonadectomy of dogs and cats. *J Am Vet Med Assoc* 231:1665–1675.

23

When is it best to perform ovariohysterectomy of female dogs relative to heat?

General information

Vascularity of the ovaries and the uterus is increased when dogs are under the influence of estrogen, as in proestrus and estrus, and during pregnancy. Vascularity is decreased during a nonpregnant diestrus, when the bitch is under the influence of progesterone. Most bitches are best spayed at least 2 months after standing heat, when in anestrus.

Clinical implications

Because of increased vascularity, anesthesia and surgery take longer and there is a greater risk of intra- and postoperative bleeding if surgery is performed during proestrus or estrus. Similar risks, along with dehydration and anemia, are present if ovariohysterectomy (OHE) is performed during pregnancy, as the heavily gravid uterus is removed surgically. The normal physiologic changes of pregnancy may also complicate anesthesia and surgery (see Chapter 33). There are no published studies documenting increased incidence of side effects relative to stage of pregnancy during which OHE is performed.

The most commonly reported complication of OHE during nonpregnant diestrus is false pregnancy. False pregnancy is a misnomer; the condition is more correctly termed pseudocyesis or false whelping. Clinical signs appear concurrent with a decrease in serum progesterone concentration at the end of diestrus. OHE during diestrus, with removal of the ovaries and subsequent decrease in progesterone, mimics this normal decline in progesterone. Clinical signs include mammary development, lactation, mothering of inanimate objects, and possible aggression. Most clinical signs resolve without therapy. Wrapping the mammary area with an elastic bandage minimizes licking at the glands and increases mammary pressure, which inhibits prolactin production. This, coupled with cutting food and water in half for a day, may hasten resolution. Prolactin-inhibiting drugs can be used to inhibit lactation; the only such drug approved for use in dogs is cabergoline (5 μg/kg once daily for 5 to 7 days). If a sedative is needed to quiet an aggressive bitch, a benzodiazepine such as diazepam should be used in preference to a phenothiazine such as acepromazine, as the latter may stimulate prolactin release.

Research suggests that some breeds of dog may become more reactive or aggressive after OHE. Anecdote suggests that these changes are more likely if bitches are spayed during diestrus; there are no studies supporting this contention.

Supplemental reading

Johnston SD, Root Kustritz MV, Olson PNS. 2001. *Canine and Feline Theriogenology*. Philadelphia: WB Saunders Co., pp. 243–246.

Romagnoli S. 2008. Surgical gonadectomy in the bitch and queen: Should it be done and at what age? Proceedings, Southern European Veterinary Conference, Barcelona, Spain.

24

What age is best for castration of male dogs?

General information

While studies have demonstrated the safety of castration in dogs as young as 7 weeks of age, there are no prospective long-term studies demonstrating the optimal age for castration of male dogs. Non-owned dogs (stray or feral animals, humane society animals) should be castrated prior to placement in a new home. Owned animals should be assessed on a case-by-case basis, weighing the pros and cons. There is much in the popular literature on this topic; veterinarians are cautioned to be aware of the number and validity of studies cited to support statements of fact in any publication.

Pros of castration include a decrease in incidence of sexually dimorphic behaviors, and decreased incidence of diseases of the prostate and testes. The sexually dimorphic behaviors best controlled by castration are mounting, roaming, and urine marking. Aggression is often not controlled by castration, as the underlying cause of aggression is usually not hormonal.

Testicular neoplasia is very common in dogs, with a reported incidence of 0.9%. Testicular tumors usually arise in aged dogs and are rarely malignant; castration is curative.

Prostate disease is very common in intact male dogs. By 6 years of age, 75 to 80% of intact male dogs have histologic or clinical evidence of benign prostatic hypertrophy (BPH)(see Chapter 53). BPH is usually not associated with severe clinical signs but does predispose the dog to prostatitis, which may be associated with substantial morbidity (see Chapter 54). Castration is a cure for BPH and promotes resolution of prostatitis.

Cons of castration include weight gain, predisposition to cranial cruciate ligament (CCL) injury, and possible predisposition to several uncommon, high morbidity cancers including prostatic adenocarcinoma, osteosarcoma, and hemangiosarcoma.

Obesity is the most common nutritional disorder in dogs and multiple retrospective studies have demonstrated a correlation between castration and weight gain. In cats, it has been demonstrated that gonadectomy is associated with decreased metabolic rate and it may well be that the same is true in dogs. Weight gain is easily controlled by the owner, with appropriate diet and exercise.

CCL injury is more common in women than in men and more commonly occurs during specific stages of the menstrual cycle, suggesting that hormonal factors may play a role. Several studies have demonstrated increased incidence of CCL injury in spayed or neutered dogs, compared with intact dogs; obesity was not indicated as a contributing factor in these studies. Studies have not demonstrated asynchrony in bone growth secondary to gonadectomy, or excessive angulation of the tibial plateau as causative factors.

Prostatic adenocarcinoma is an uncommon, highly malignant tumor of castrated or intact male dogs. Several studies have demonstrated two to four times increased incidence in castrated males compared with age-matched intact male dogs. Similarly, increased incidence of osteosarcoma, a highly malignant tumor with a reported incidence of 0.2%, was reported in castrated populations compared with intact populations in two studies, one of which evaluated the disease in Rottweilers, which have a hereditary predisposition to osteosarcoma. Hemangiosarcoma is another highly malignant, low-incidence tumor reported to have increased incidence in castrated dog populations in two studies. No cause-and-effect relationship has been defined for any of these tumors.

Clinical implications

Veterinarian and clients must take into account incidence, and morbidity and mortality of disorders associated with castration to determine optimal age of castration for a given dog. Breed may also play a role. The author strongly encourages complete evaluation of likelihood that a male will be allowed to breed indiscriminately; if odds are high that reproduction will be controlled, there are few compelling medical reasons to castrate the animal when young.

Supplemental reading

Root Kustritz MV. 2007. Determining the optimal age for gonadectomy of dogs and cats. *J Am Vet Med Assoc* 231:1665–1675.

25

Are nonsurgical alternatives available for sterilization or contraception of female or male dogs?

General information

Contraception is reversible control of reproduction; sterilization is nonreversible or permanent control of reproduction. In companion animal medicine, most often we are interested in the latter. As of this writing, there are no nonsurgical sterilants commercially available that are completely effective and safe for use in dogs and cats. This discussion will primarily concern nonsurgical contraceptives.

Females

Hormones

Progesterone

For female dogs, a synthetic form of progesterone, megestrol acetate (Ovaban™) was commercially available for many years and is still an approved estrus-suppressing drug for breeding and nonbreeding bitches. It is not available commercially at this time but can be dispensed as a human brand of megestrol acetate through a human pharmacy. Megestrol acetate works by exerting negative feedback to the pituitary, preventing release of follicle stimulating hormone (FSH) and luteinizing hormone (LH). The drug can be given in anestrus for 30 days (0.55 mg/kg [0.25 mg/lb] once daily per os) or within the first 3 days of proestrus, for 8 days (2.2 mg/kg [1 mg/lb] once daily per os). If the drug is given in proestrus, the bitch will continue to show physical and behavioral signs of heat for at least 5 days and may permit mounting and breeding by the male. If the drug is administered for 3 consecutive days before breeding occurs, she most likely will not ovulate and conceive on that cycle. After treatment in either anestrus or proestrus, average return to proestrus is in 4 to 6 months. There is a wide range of possible duration from treatment to next proestrus; this drug cannot be used to suppress a heat cycle and make it occur at a time more desired by the owner. The manufacturer recommends that this drug not be used to suppress a bitch's first heat and that it not be used to suppress more than two consecutive heat cycles. Side effects in bitches include increased appetite and mammary development. Progestins should not be administered to bitches with a history of mammary neoplasia, pyometra, or diabetes mellitus.

Many other forms of progesterone are available as contraceptives in other countries. These include injectable forms of progesterone, such as medroxyprogesterone acetate, and unique progestin compounds reported to have minimal side effects, such as proligestone. In many countries where performance of elective ovariohysterectomy (OHE) and castration is unethical or illegal, progestins are routinely administered as injections or via a patch, implant, or other slow-release systems. These compounds are not available in the United States.

Androgen

A mild androgen, mibolerone (Cheque™) is approved for use in nonbreeding bitches in the United States. This drug is not approved for use in breeding bitches. Cheque™ is no longer available commercially but mibolerone can be procured through compounding pharmacies. Androgens inhibit estrus by exerting negative feedback on the pituitary. Mibolerone dose is dependent on the weight of the dog and breed; dogs weighing 0.5 to 12 kg receive 30 μg once daily per os, those weighing 12–23 kg 60 μg once daily per os, those weighing 23–45 kg 120 μg, and those weighing over 45 kg and German Shepherd purebred or crosses 180 μg. The drug is administered once daily per os for as long as the dog is to be kept out of heat. Therapy must be instituted at least 30 days before the next expected proestrus to be effective. After the drug is withdrawn, average duration to next proestrus is 70 days, with a wide range of normal. Reported side effects include clitoral hypertrophy, musky body odor, exudation of creamy vulvar discharge, increased mounting and aggressive behaviors, obesity, and epiphora.

In the United States, testosterone is not approved for estrus suppression in bitches, although there are reports of efficacy of testosterone conjugates as estrus-suppressing drugs in bitches. Future fertility may be altered by administration of this strong androgen.

Gonadotropin releasing hormone (GnRH) agonists and antagonists

GnRH agonists initially induce estrus, then suppress estrus through long-term down-regulation of FSH and LH release. Most are administered via an osmotic pump or implant. No products using these drugs are approved for use in bitches or are commercially available in the United States.

GnRH antagonists act by blocking release of GnRH and subsequently decreasing secretion of FSH and LH. Use of this class of drug early in proestrus was associated with lack of ovulation on that cycle and return to proestrus in 3 weeks. No products using these drugs are approved for use in bitches or are commercially available in the United States.

Immunocontraception

Immunocontraception relies on the body's immune response to block fertility by stimulating creation of antibodies against specific molecules or tissues. The two antigens best described are zona pellucida (ZP) proteins, which encircle the egg, and GnRH.

ZP proteins are highly conserved between species. This permits use of readily available pig ovaries as a source of antigen. At present, there is no commercially available ZP vaccine for dogs. Research has documented decreased fertility after immunization with ZP proteins; however, boostering of vaccine is required and some bitches will continue to exhibit physical and behavioral signs of estrus, often at unpredictable intervals, making this solution unacceptable to pet owners.

GnRH is a small molecule and, as such, is not inherently antigenic. Conjugation of GnRH with larger proteins enhances antigenicity. At present, there is no commercially available GnRH vaccine for dogs. However, a vaccine for hoofstock is undergoing review for approval and similar products may be available for companion animals in the future.

Mechanical/physical barriers

Examples of mechanical devices that have been described in dogs include an intravaginal contraceptive device and an intrauterine device. The high failure rate of the intravaginal device led to its being removed from the market decades ago. Intrauterine devices may or may not be impregnated with spermicidal compounds; these devices suffer from difficulty in placement, and the likelihood that bitches will continue to demonstrate estrous cycling.

Males

Hormones

FSH and LH are required for normal male fertility. The negative feedback effect of progestins is not enough to significantly impair spermatogenesis in male dogs. Similarly, administration of testosterone and androgen antagonists has not been demonstrated to impair spermatogenesis. GnRH analogs have been demonstrated to induce reversible infertility by decreasing LH and testosterone concentrations with subsequent decline in spermatogenesis. While there are reports of use of these drugs in dogs, their use is best described in zoo animals and no products using this class of drugs are approved for use in dogs in the United States.

Immunocontraception

Antigens used to induce antibodies against molecules or tissues of reproductive interest in male dogs include spermatozoal antigens and GnRH. There are no reports of success with spermatozoal antigens in dogs. GnRH is a small molecule and, as such, is not inherently antigenic. Conjugation of GnRH with larger proteins enhances antigenicity. At present, there is no commercially available GnRH vaccine for dogs. However, a vaccine for hoofstock is undergoing review for approval and similar products may be available for companion animals in the future.

Mechanical/physical barriers

Injection of agents into the epididymis, testes, or vasa deferentes has been described. These compounds are irritating, causing localized inflammation and either destroying tissue locally or creating scar tissue and impeding flow of spermatozoa and seminal fluid for ejaculation. While several compounds have been described for this use, only one, zinc gluconate buffered with arginine (Neutersol™ or Esterisol™), has been approved for use in dogs in the United States. This compound is administered in each testis, with dose dependent on testicular width. Neutersol™ was approved for use in dogs aged 3 to 10 months of age and was not to be used in dogs with misshapen or retained testes. Side effects included transient scrotal swelling and discomfort, and vomiting. The majority of dogs tested either had no ejaculate or an ejaculate containing no spermatozoa by 6 to 12 months after administration. Serum testosterone concentration in treated dogs was half of that in untreated dogs. As of this writing, no long-term independent studies have been published using either product.

Clinical implications

Most pet owners are as concerned with controlling their bitch's or dog's physical and behavioral manifestations of reproduction as they are with controlling reproduction itself. Current nonsurgical methods either are not efficacious in controlling reproduction or permit continuing physical changes of heat and normal reproductive behaviors that owners consider objectionable. Surgery (OHE or castration) is the standard for reproduction control in the United States at this time (see Chapters 22 to 24).

Supplemental reading

Bowen RA. 2008. Male contraceptive technology for nonhuman male mammals. *Anim Reprod Sci* 15:139–143.

Jana K, Samanta PK. 2007. Sterilization of male stray dogs with a single intratesticular injection of calcium chloride: A dose-dependent study. *Contraception* 75:390–400.

Kutzler M, Wood A. 2006. Non-surgical methods of contraception and sterilization. *Theriogenology* 66:514–525.

Walker J, Ghosh S, Pagnon J, et al. 2007. Totally synthetic peptide-based immunocontraceptive vaccines show activity in dogs of different breeds. *Vaccine* 25:7111–7119.

26

What comprises recommended pre-breeding evaluation for bitches?

General information

The goal in breeding is always to create offspring that are superior to the dam and sire. This requires assessment of health, infectious disease, and genetic predispositions carried by the dam.

Health assessment requires a complete physical examination and performance of a digital vaginal examination. Vaccinations should be current and the bitch should be checked and treated for internal and external parasites. For infectious disease, serologic testing for canine brucellosis should be performed (see Chapter 44). The occasional stud owner will require vaginal culture of the bitch. The vagina is not a sterile surface and to ask for a negative vaginal culture is not appropriate. If culture is performed, it is best done when the bitch is in proestrus or estrus as vaginal culture will permit assessment of the discharge of heat, which arises in the uterus.

Hereditary diseases of concern vary by breed. The author strongly recommends that breeders access information from the health committee of their national breed club to learn which conditions are of greatest concern in their breed and what tests are available.

Types of testing available include anatomic testing and DNA-based testing. Anatomic testing is evaluation of a morphologic characteristic, such as degree of laxity in the hip joint (University of Pennsylvania Hip Improvement Program) or presence of subluxation and osteophytes in the hip joint (Orthopedic Foundation for Animals hip radiographs). Anatomic tests may suffer from subjectivity. DNA-based tests can be either direct tests, which identify specific mutations on the animal's DNA predisposing them to disease or designating them a carrier, or linkage tests, which identify DNA markers near the site of an affected gene. Direct tests are more objective but linkage tests can be quite accurate. New DNA-based tests become available constantly and again, breeders are encouraged to work with their national breed club to keep abreast of concerns in their breed.

Clinical implications

Testing for canine brucellosis well before onset of proestrus permits repeated testing if screening tests are positive (see Chapter 44). However, some authors feel that seropositive bitches are more likely to be identified if they are tested when in proestrus or estrus.

For vaginal culture, a mixed growth of aerobic bacteria and *Mycoplasma* sp. is normal. A significant culture result is moderate to heavy growth of any single organism. Mycoplasma organisms are hard to grow and many laboratories will not give you an idea of quantity grown, making interpretation of culture results difficult. Bitches with no history of infertility and no signs of reproductive tract disease who culture positive for Mycoplasma probably do not require treatment (see Chapter 28).

Breeders will come for anatomic tests, such as hip radiographs. Some tests must be performed by a licensed specialist; for example, Canine Eye Registry Foundation examinations must be completed by a boarded veterinary ophthalmologist. DNA-based tests are usually performed on cheek swabs, skin biopsies, or other readily obtained samples; breeders may or may not request assistance with sample collection or interpretation of results.

Supplemental reading

Johnston SD, Root Kustritz MV, Olson PNS. 2001. *Canine and Feline Theriogenology*. Philadelphia: WB Saunders Co., pp. 41–46.

27

What effect does hypothyroidism have on fertility?

General information

Hypothyroidism is the most common endocrine disorder in dogs. Most cases are due to autoimmune thyroiditis and the inciting cause is rarely identified; vaccination is not associated with autoimmune thyroiditis in dogs. Diagnosis should not be made by assay of thyroxine (T4) alone as values vary with breed, stage of the estrous cycle, concurrent drug therapy, and health status. Accurate diagnosis requires demonstration of decreased serum concentrations of free T4 and increased concentrations of thyroid stimulating hormone.

There are reports and anecdotal evidence linking hypothyroidism with abnormal estrous cycles (altered frequency and duration), primary anestrus, conception failure, galactorrhea, and pregnancy loss in bitches. However, in one study of 18 bitches with experimentally induced hypothyroidism, there was no difference between the treatment and control groups in interestrous interval, litter size, or gestation length. Treated bitches showed slower uterine contractions with subsequent prolonged labor, and increased periparturient puppy mortality.

Hypothyroidism has been anecdotally associated with decreased libido, poor semen quality, and retrograde ejaculation in dogs. However, no study has demonstrated a cause-and-effect relationship between hypothyroidism and fertility in male dogs.

Because hypothyroidism is caused by autoimmune destruction of the thyroid gland with subsequent atrophy, it is possible that concurrent autoimmune destruction of the thyroid and gonads may be responsible for this apparent link. Several studies in male dogs suggest this connection. This does not explain why some hypothyroid dogs with poor reproductive function return to normal fertility when properly supplemented with thyroid hormone.

Clinical implications

Hypothyroidism has a hereditary component. Although mode of inheritance is not well understood, the increased incidence in certain breeds suggests a genetic predisposition. It may be most valuable to identify hypothyroidism in breeding stock to help the breeder decide if that individual should remain in the gene pool.

If a dog is believed to have poor reproductive function secondary to hypothyroidism, accurate diagnosis and appropriate testing after supplementation is instituted are required. Unnecessary treatment with thyroid hormone will induce iatrogenic hypothyroidism when the drug is withdrawn. Over-supplementation and creation of a hyperthyroid state is not beneficial to the dog's general health and may be associated with reproductive dysfunction as is seen in hyperthyroid humans.

Supplemental reading

Ferguson DC. 2007. Testing for hypothyroidism in dogs. *Vet Clin North Am* 37:647–669.

Johnson CA. 2002. Thyroid issues in reproduction. *Clin Tech Small Anim Pract* 17:129–132.

Panciera DL, Purswell BJ, Kolster KA. 2007. Effect of short-term hypothyroidism on reproduction in the bitch. *Theriogenology* 68:316–321.

Scott-Moncrieff JC, Glickman NW, Glickman LT, et al. 2006. Lack of association between repeated vaccination and thyroiditis in laboratory Beagles. *J Vet Intern Med* 20:818–821.

Shiel RE, Brennan SF, Omodo-Eluk AJ, et al. 2007. Thyroid hormone concentrations in young, healthy, pretraining Greyhounds. *Vet Rec* 161:616–619.

Torres SMF, Feeney DA, Lekcharoensuk C, et al. 2003. Comparison of colloid, thyroid follicular epithelium, and thyroid hormone concentrations in healthy and severely sick dogs. *J Am Vet Med Assoc* 222:1079–1085.

Wood MA, Panciera DL, Berry SH, et al. 2008. Influence of isoflurane general anesthesia or anesthesia and surgery on thyroid function tests in dogs. *J Vet Intern Med.* doi:10.1111/j.1939-1676.2008.0216.x.

28

What are the pros and cons of treatment with antibiotics during heat in bitches?

General information

The owner of a bitch with a history of infertility or small litter size may ask for antibiotics to be dispensed for administration during heat and pregnancy as an empirical treatment for disease. Bitches with a history of pyometra must be treated with antibiotics during heat to control secondary infection. The occasional stud owner will require vaginal cultures of the bitch as part of a pre-breeding workup. The vagina is not a sterile surface and to ask for a negative vaginal culture is not appropriate.

The class of antibiotic used during heat, pregnancy, and lactation must be one that does not interfere with fertilization, implantation, embryonic development, fetal growth, and normal physiologic function of the nursing pups. Most information in the veterinary literature refers to teratogenicity of drugs used (see Chapter 30). In general, tetracyclines, aminoglycosides, and chloramphenicol should be avoided, and fluoroquninolones should be used with caution.

Studies have documented the pros and cons of use of empirical antibiotics during heat in bitches with a history of reproductive tract disease. Older bitches are more likely to conceive and to carry pregnancy if treated with antibiotics during heat. Cons include creation of resistant bacterial strains from indiscriminate antibiotic use and possible overgrowth of normal flora, including *Mycoplasma* sp., after antibiotic therapy alters the normal balance of the vaginal flora.

Clinical implications

If antibiotics are to be administered, it is always in the animal's best interest to perform culture and sensitivity testing to guide antibiotic choice. If culture is performed, it is best done when the bitch is in heat as vaginal culture will permit assessment of the heat discharge, which arises in the uterus. It has been demonstrated that a larger number of organisms are cultured when samples are collected during estrus than when samples are collected during diestrus or anestrus. It has also been demonstrated that fewer contaminants will be present in samples collected from the cranial vagina than from the caudal vagina. Use of a long, guarded swab decreases contamination.

For vaginal culture, a mixed growth of aerobic bacteria and *Mycoplasma* sp. is normal. A significant culture result is moderate to heavy growth of any single organism. Mycoplasma organisms are hard to grow and many laboratories will not give an idea of quantity grown, making interpretation of culture results difficult. Bitches with no history of infertility and no signs of reproductive tract disease who culture positive for Mycoplasma probably do not require treatment.

Supplemental reading

Bjurstrom L, Linde-Forsberg C. 1992. Long-term study of aerobic bacteria of the genital tract in breeding bitches. *Am J Vet Res* 53:665–673.

Root Kustritz MV. 2006. Collection of tissue and culture samples from the canine reproductive tract. *Theriogenology* 66:567–574.

29

How do I use progesterone and other measures for breeding management?

General information

Progesterone assay

Serum progesterone concentrations begin to rise before ovulation in bitches, with a significant rise from baseline values at about the time of peak secretion of luteinizing hormone (LH) from the pituitary. On average, serum progesterone concentration will be in the range of about 2.0 to 2.9 ng/ml 24 to 48 h prior to ovulation, and will be anywhere from 4.0 to 10.0 ng/ml on ovulation day. Blood samples submitted to commercial laboratories for progesterone assay will yield accurate, quantitative results either in ng/ml or nmol/l (ng/ml × 3.1 = nmol/l). Blood samples evaluated with in-house test kits are much less quantitative, usually identifying concentrations as either below or above 4 to 5 ng/ml. Because of demonstrated variability in changes in progesterone around the time of ovulation between and within bitches, practitioners are strongly advised to continue to monitor progesterone concentrations until they are well within or beyond the ovulatory range. There is no consistent single value of serum progesterone concentration associated with optimal breeding day (Table 29-1).

Luteinizing hormone assay

LH is the hormone that stimulates ovulation. It is secreted as a single large peak, with high concentrations persisting for about 24 to 48 h. Serum concentrations of greater than 1 ng/ml precede ovulation by 2 to 3 days. Commercial assays for LH are available but turnaround time is so long as to preclude their use for breeding management. The only commercially available in-house LH assay varies in its availability. No one has identified a consistent temperature change at the time of maximum LH release in bitches, as has been identified in women; this may be due to the subtlety of the effect.

Vaginal cytology

The population of vaginal epithelial cells collected by vaginal swab varies in a characteristic manner throughout the estrous cycle (see Chapters 1 and 2). The average bitch ovulates about 48h after onset of estrus, defined cytologically as 100% cornification with greater than 50% being anuclear squame cells. However, great variability exists between and within bitches, and vaginal cytology alone cannot be used accurately to prospectively determine ovulation day in bitches. In one study, only 10 of 35 bitches (28%) would have been bred at the right time if vaginal cytology

Table 29-1. Use of serum progesterone concentrations for breeding management in bitches.

Serum Progesterone Concentration (ng/ml)	Interpretation
Less than 1.0	Anestrus or early proestrus—recheck.
1.0 to 2.0	LH peak has not yet occurred—recheck.
2.0 to 4.0	LH peak occurring, near ovulation—recheck.
4.0 to 10.0	Ovulation occurring.
Greater than 10.0	Ovulation has occurred.[a]

[a]Vaginal cytology must be used to determine if bitch has ovulated and is still in estrus or if bitch has ovulated and has gone into diestrus (see Chapter 2 and Clinical Implications).
LH = luteinizing hormone.

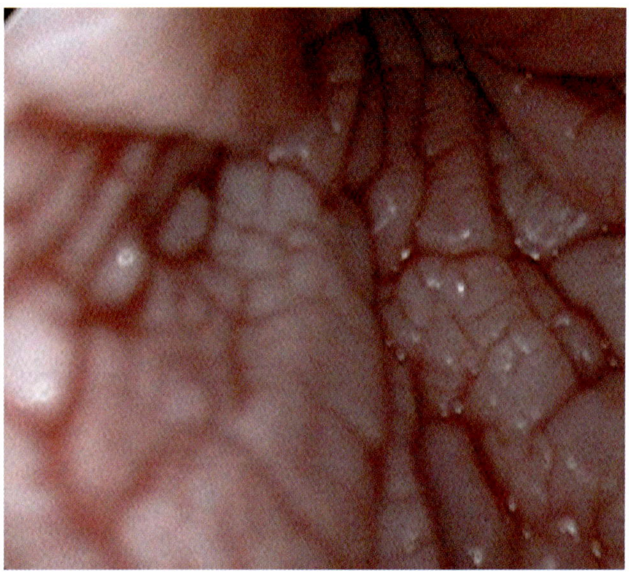

Figure 29-1: Blanched and crenulated vaginal mucosa associated with estrus, as viewed through a vaginoscope.

had been the only measure used for breeding management. In a study of 292 estrous cycles, ovulation day averaged 11 to 12 days after proestrus onset but ranged from 1 to 31 days.

Vaginoscopy

The appearance of the vaginal mucosa changes in a characteristic way from proestrus into estrus (see Chapter 3). The average bitch ovulates about 48h after first appearance of the maximum crenulation and blanching associated with onset of estrus (Fig. 29-1). However, great variability exists between and within bitches, and vaginoscopy alone cannot be used accurately to prospectively determine ovulation day in bitches.

Behavioral and physical changes

The average bitch ovulates about 48 h after onset of estrus, defined behaviorally as first acceptance of the male dog. Oftentimes, the vulva becomes less turgid and vulvar discharge of heat changes from blood-tinged to straw-colored concurrent with onset of estrus as well. Great variability exists between and within bitches, and neither behavioral changes of the bitch or dog, changes in vulvar size or tone, or changes in appearance of the vulvar discharge can be used accurately to prospectively determine ovulation day in bitches.

Miscellaneous

Ultrasound evaluation of ovaries

Reports exist documenting changes in ovarian appearance at the time of ovulation. These changes can be identified using abdominal ultrasound. However, the technique requires significant training and very high quality ultrasound equipment, and is not commonly used in practice.

Electrical impedance

Resistance of electrical conductivity across the vaginal mucosa may increase coincident with increasing estrogen concentrations. Although one may purchase equipment that accurately measures electrical conductivity across the vaginal mucosa, there are no published reports demonstrating a correlation between electrical impedance measurements and occurrence of the LH peak or ovulation in bitches.

Glucose concentration in vaginal fluid

Anecdotal reports suggest that glucose concentration rises in vaginal fluid coincident with the rise in serum progesterone that occurs around the time of ovulation. There are no published reports documenting this as an accurate technique for determination of ovulation day.

Ferning of vaginal fluid

Fluid from the cranial vagina allowed to spread across a glass slide and air-dry will take on the branched appearance of a fern, with the effect most pronounced around the time of the LH surge. Ferning has not been demonstrated to consistently permit identification of ovulation day in bitches.

Clinical implications

The only diagnostic test readily and consistently available to practitioners at this time for determination of ovulation day in bitches is measurement of serum progesterone concentration. Serum progesterone should be measured until it is well into the ovulatory range. Lower values, while indicative of significant events in the average dog, are not consistent enough in all bitches to discourage serial progesterone assay. After ovulation occurs, serum progesterone concentrations remain high for 2 months, throughout diestrus. There is great variation in how quickly serum progesterone rises after ovulation. Vaginal cytology, which is completely cornified throughout estrus, will abruptly become completely non-cornified on the first day of diestrus; this change occurs with fair consistency 6 days after ovulation (see Chapter 2). Therefore, if a bitch has a serum progesterone concentration greater than 10 ng/ml and still has cornified cytology, she is somewhere in that 6-day window and may even be at optimal breeding day. However, if a bitch has a serum progesterone concentration greater than 10 ng/ml and has non-cornified cytology, she is in diestrus and cannot be bred this season.

Serum progesterone concentration can be altered by sample handling. Practitioners are strongly encouraged to call the laboratory to which they will be submitting samples to ask for specific information regarding the type of tube into which blood should be collected, how soon after blood collection the sample should be centrifuged or otherwise processed, and how serum or plasma should be stored and shipped.

Because there is variability in serum progesterone concentrations in bitches, it is useful to become familiar with one other technique, not to help pinpoint ovulation day but perhaps to ensure any overt abnormalities with the reproductive cycle in that bitch are identified and addressed in a timely manner. The author routinely uses serum progesterone concentration and vaginal cytology; others use serum progesterone and vaginoscopy. Readers are encouraged to use whatever technique they are most confident in interpreting.

Timing of insemination after ovulation varies with type of semen used. If only one breeding is possible, optimal breeding day is 2 days after ovulation (4 days after the LH peak) for natural service, fresh semen or chilled semen (see Chapter 12). Conception rate and litter size are higher if the bitch is bred more than once; many recommend breeding 2 and 4 days after ovulation. If frozen-thawed semen is used, optimal breeding day is later because spermatozoa viability is reduced. Insemination is recommended 3 to 5 days after ovulation (5 to 7 days after the LH peak).

The occasional bitch will exhibit physical and cytologic changes of proestrus with no concomitant rise in progesterone, will appear to go out of heat, and then will go through a normal ovulatory heat about 1 month later. This is termed a split heat and was reported to occur in about 1.2% of dogs in one large study.

Supplemental reading

Arbeiter K. 1993. Anovulatory ovarian cycles in dogs. *J Reprod Fertil* 47:453–456.

Fay J, Mezo T, Solti L, et al. 2003. Comparison of different methods used for oestrus examination in the bitch. *Acta Vet Hung* 51:385–394.

Hewitt D, England G. 2000. Assessment of optimal mating time in the bitch. *In Prac* 22:24–33.

Hiemstra M, Schaefers-Okkens AC, Teske E, et al. 2001. The reliability of vaginal cytology in determining the optimal mating time in the bitch. *Tijdschr Diergeneeskd* 126:685–689.

Johnston SD, Root Kustritz MV, Olson PNS. 2001. *Canine and Feline Theriogenology*. Philadelphia: WB Saunders Co., pp. 49–55.

Nishiyama T, Narita K, Tsumagari S, et al. 2000. Shrinkage in the horizontal dimensions of the vulva (vulvar shrinkage) as an indicator of standing heat in the Beagle. *J Am Anim Hosp Assoc* 36:556–560.

Root Kustritz MV. 2006. *The Dog Breeder's Guide to Successful Breeding and Health Management*. St. Louis, MO: Elsevier, pp. 135–140.

Thuroczky J, Wolfling A, Tibold A, et al. 2003. Effect of anticoagulants and sampling time on results of progesterone determination in canine blood samples. *Reprod Dom Anim* 38:386–389.

Tsumagari S, Ichikawa Y, Toriumi H, et al. 2003. Optimal timing for canine artificial insemination with frozen semen and parentage testing by microsatellite markers in superfecundency. *J Vet Med Sci* 65:1003–1005.

30

What drugs are unsafe to use during pregnancy in bitches?

General information

Drugs can be classified using the following designation: A: probably safe in dogs, based on the lack of toxicity or teratogenicity identified in other species; B: safe if used cautiously; C: potentially risky and to be used only when the benefit clearly outweighs the risks; and D: contraindicated in all cases. Few drug studies have been completed in bitches, requiring extrapolation from work in other species, most commonly humans and laboratory animals (Table 30-1).

Clinical implications

All kinds of drug use during pregnancy require careful weighing of benefits and detriments. Because of the lack of research in bitches and subsequent need to extrapolate from other literature, good client communication regarding extra-label use of drugs is critical.

Supplemental reading

Budsberg SC. 2009. Review of NSAIDs: COX selectivity and systemic effects beyond analgesia. *Compendium* 31 (Suppl 3A):10–16.

Johnston SD, Root Kustritz MV, Olson PNS. 2001. *Canine and Feline Theriogenology*. Philadelphia: WB Saunders Co., pp. 81–86.

Landsbergen N, Pelicaan CH, Schaefers-Okkens AC. 2001. The use of veterinary durgs during pregnancy of the dog. *Tijdschr Diergeneeskd* 126:716–722.

Novotny MJ, Krautmann MJ, Ehrhart JC, et al. 2000. Safety of selamectin in dogs. *Vet Parasitol* 91:377–391.

Root Kustritz MV. 2006. *The Dog Breeder's Guide to Successful Breeding and Health Management*. St. Louis, MO: Elsevier, pp. 21–45.

Traas AM. 2008. Surgical management of canine and feline dystocia. *Theriogenology* 70:337–342.

Table 30-1. Safety of drugs for use in pregnancy.

Chemical Name	Safety Designation[a]	Comments
Antibiotics		
Ampicillin	A	—
Amoxicillin	A	—
Amoxicillin-clavulanate	A	—
Cephalosporins	A	—
Chloramphenicol	C	May decrease protein synthesis in fetus.
Doxycycline	D	All tetracyclines may cause bone and teeth malformations in fetus.
Erythromycin	A	—
Fluoroquinolones	C	All drugs in this class have been associated with cartilage damage; however, multiple authors have reported no birth defects in pups born to bitches treated with fluoroquinolones.
Gentamicin	C	Aminoglycosides are associated with ototoxicity and nephrotoxicity.
Metronidazole	C	Teratogenic in laboratory animals.
Trimethoprim-sulfa	B	—
Pre-anesthetics, anesthetics, sedatives		
Acepromazine	B	May cause fetal CNS depression.
Atropine	B	May cause fetal tachycardia.
Diazepam	C	—
Glycopyrrolate	B	—
Isoflurane	B	—
Ketamine	B	May induce premature labor.
Naloxone	A	—
Oxymorphone	B	—
Pentobarbital	C	May cause respiratory depression in fetus.
Propofol	C	Avoid repeated injections, as drug may accumulate within the fetus due to inability to clear it readily.
Sevoflurane	B	—
Thiopental	C	May cause respiratory depression in fetus.
Parasiticides		
Diethylcarbamazine	A	—

Table 30-1. *Continued*

Chemical Name	Safety Designation[a]	Comments
Fenbendazole	A	—
Ivermectin	A	—
Milbemycin oxime	A	—
Piperazine	A	—
Praziquantel	A	—
Pyrantel	A	—
Selamectin	A	—
Thiabendazole	B	High doses cause toxemia in ewes.
Pain medications		
Acetaminophen	C	—
Aspirin	C	Embryotoxicity in laboratory animal species, possible pulmonary hypertension and bleeding in late pregnancy.
Butorphanol	C	—
Carprofen	C	Increased gestation length and increase in stillbirths identified in laboratory animals.
Flunixin meglumine	C	—
Ibuprofen	C	—
Selective cyclooxygenase-2 inhibitors	D	In humans, ingestion of drugs of this class may retard renal development in neonates.
Endocrine steroids		
Dexamethasone	C	Increased incidence of cleft palate and other congenital defects; may induce premature labor. High doses at mid-gestation may cause pregnancy termination.
Diethylstilbestrol	D	May induce abnormal development of reproductive system.
Prednisolone	C	—
Thyroxine	B	—

[a]A, probably safe in dogs, based on the lack of toxicity or teratogenicity identified in other species; B, safe if used cautiously; C, potentially risky and to be used only when the benefit clearly outweighs the risks; D, contraindicated in all cases.

31

Are there dietary supplements proven to improve fertility in bitches?

General information

General nutrition requirements provided on pet food labels and promoted by the Association of American Food Control Officials (AAFCO) reflect minimum requirements and are not a reflection of optimum percentages of protein, fat, carbohydrates, and macro- and micronutrients. These percentages are not described in the veterinary literature. Clients should be encouraged to feed their bitches with a commercial food labeled by AAFCO as nutritionally complete and balanced for pregnancy and lactation, or for all life stages.

Calcium supplementation is often requested by clients, with the thought that the bitch will somehow build up calcium reserves that will be depleted during lactation. Unfortunately, calcium supplementation during gestation may down-regulate the normal calcium regulatory mechanism and there is no evidence to suggest that elevated concentrations in bone result from oral supplementation during pregnancy. Commercial diets should have a calcium:phosphorus ratio of 1:1 to 1.2:1. Commercial diets are recommended during pregnancy to ensure this ratio throughout gestation; this may be difficult for bitches fed homemade diets, which are more likely to contain widely varying ingredients over time and are unlikely to be evaluated for nutrient content.

Prebiotics are nutritional supplements that support growth and function of intestinal microflora. Probiotics contain active bacteria that can recolonize the intestinal tract and promote recolonization of the normal mixed flora. Purchasers of probiotics are well advised to look for products that are formulated to ensure viability through processing, storage, and passage through the digestive system. These compounds may be useful as adjunct therapy in bitches with physiologic diarrhea during pregnancy or lactation, or with diarrhea secondary to antibiotic therapy.

Fatty acid deficiency in diets fed to pregnant bitches may be associated with preterm labor, poor placental development, and decreased litter size. Omega-3 fatty acids, as may be found in fish oil or flaxseed preparations, are important for normal neurological development of puppies. These compounds, provided to bitches during estrus, pregnancy, and lactation, have not been demonstrated to enhance fertility but have been demonstrated to stimulate neurologic development of the pups, as identified by enhanced electroretinographic responses. Because bitches show variable ability to metabolize fatty acids based on initial nutrient source, specific recommendations for concentrations to be fed are often not provided.

Supplements described as fertility-enhancing that contain herbs, vitamins, and amino acids are available commercially. Some of these compounds have been demonstrated to have effects in species other than the dog. Folate has been demonstrated to be important for oocyte quality and for maturation of the conceptus, implantation and placentation, and fetal growth in humans. Zinc may play a role in ovulation in women. Both folate and zinc have antioxidant properties in humans, in whom it has been demonstrated that oxidative stress plays a role in subfertility and infertility. Vitamin E also has antioxidant activities and has been shown in humans to modulate gene expression. However, excessive vitamin E supplementation may be associated with creation of inflammation. A human proprietary product containing chasteberry, green tea, L-arginine, vitamins including folate, and minerals, was demonstrated to normalize cycle length in women with either abnormally short or long cycles and to increase conception rate (26% in the treatment group vs. 10% in the placebo group; $n = 93$).

Clinical implications

General recommendations for feeding during pregnancy include feeding a commercial diet with AAFCO approval that is labeled as nutritionally complete and balanced for pregnancy or for all life stages. Do not supplement with calcium during pregnancy; when the bitch begins lactation, she will not be able to take in enough calcium orally to maintain milk production and will have to draw from supplies in the bone. Because the system may have been down-regulated, there will be lag time before parathormone release and subsequent osteoclastic activity can increase serum calcium levels. The bitch may actually be more likely to suffer clinical hypocalcemia during this period than if she had not been supplemented with calcium. Pre- and probiotics may be beneficial in pregnant bitches with diarrhea and are unlikely to do harm. The same can be said of some commercial fertility-enhancing products but many contain a mix of compounds in varying and often unsubstantiated concentrations, some of which may not be beneficial and may, in fact, be detrimental.

Supplemental reading

Agarwal A, Gupta S, Sharma RK. 2005. Role of oxidative stress in female reproduction. *Reprod Biol Endocrin* 3:28–49.

Bauer JE. 2001. Evaluation of nutraceuticals, dietary supplements, and functional food ingredients for companion animals. *J Am Vet Med Assoc* 218:1755–1760.

Bauer JE. 2007. Responses of dogs to dietary omega-3 fatty acids. *J Am Vet Med Assoc* 231:1657–1661.

Ebisch IMW, Thomas CMG, Peters WHM, et al. 2007. The importance of folate, zinc and antioxidants in the pathogenesis and prevention of subfertility. *Hum Reprod Update* 13:163–174.

Greco DS. 2008. Nutritional supplements for pregnant and lactating bitches. *Theriogenology* 70:393–396.

Robinson N. 2003. Herbs for pregnant animals—yes or no? *Vet Prac News* 15:28–33.

Valastyan S, Thakur V, Johnson A, et al. 2008. Novel transcriptional activities of Vitamin E: Inhibition of cholesterol biosynthesis. *Biochemistry* 47:744–752.

Westphal LM, Polan ML, Trant AS. 2006. Double-blind, placebo-controlled study of Fertilityblend: A nutritional supplement for improving fertility in women. *Clin Exp Obstet Gynecol* 33:205–208.

32

How can we prevent dystocia?

General information

Reported incidence of dystocia in dogs is 2%. Breeds predisposed to dystocia include bulldogs and other brachycephalic breeds, Scottish terriers, Chihuahuas, Dachshunds, and Yorkshire terriers.

Causes of dystocia can be maternal or fetal in origin. Maternal causes include abnormal uterine function, feto-pelvic disproportion, and pregnancy abnormalities. Examples of abnormal uterine function are primary uterine inertia, in which propulsive uterine contractions never occur in second-stage labor, and secondary uterine inertia, which is muscle fatigue associated with cessation of contractions after labor has been progressing normally. Feto-pelvic disproportion may occur when fetuses are unusually large, as in small litters, or when the pelvis is unusually small, as in bitches with a history of pelvic fracture. Pregnancy abnormalities that may be a cause of dystocia are uncommon and include uterine rupture and uterine torsion.

The most common fetal cause of dystocia is abnormal presentation. Other causes include fetal oversize, most commonly associated with small litter size, and abnormalities of fetal development (fetal monsters).

Primary uterine inertia may have a hereditary component. Abnormally weak uterine contractions may also be associated with hypocalcemia. Calcium supplementation during pregnancy is not recommended as it may be associated with down-regulation of the normal calcium regulatory mechanism. Instead, bitches should be fed a diet that is balanced for pregnancy and lactation and has a calcium:phosphorus ratio of 1.0:0.8.

Fetuses grow to fill the space available to them. Small litter size therefore predisposes bitches to dystocia. Causes of small litter size can be addressed as a means of preventing dystocia (Table 32-1).

Digital vaginal examination should be a routine component of the pre-breeding examination. Early identification of anatomic anomalies in the vagina allows complete characterization of the defect and possible surgical repair prior to the onset of proestrus (see Chapter 4).

Pelvimetry, measurement of distance and angles within the pelvis, can be performed on radiographs of the pelvis. Studies done in Boston terriers, Scottish terriers, and French bulldogs demonstrated that decreased height and width of the pelvis is associated with dystocia. Pelvic shape is heritable and varies greatly between breeds.

Table 32-1. Causes of small litter size in bitches.

Cause	Possible Solution
Improper breeding management	Bitches are optimally bred 2 days after ovulation. Ovulation date is best determined by measurement of serum progesterone concentration (see Chapter 29).
Artificial insemination (AI)	Litters sired through natural service are larger than those produced by artificial insemination, with a 15% decrease in size of litters sired using fresh or chilled semen and a 25–31% decrease in size of litters sired using frozen semen (see Chapters 12 to 16).
Age	Very young and very old dams tend to produce smaller litters. Age at which litter size is largest varies by breed, with smaller breed bitches tending to have normal-sized litters until a greater age. The "critical age" for a breed is the age at which average litter size has decreased by at least 15% from average litter size for that breed; this varies from 5 years of age in large breeds to 6 to 8 years of age in medium-sized breeds and greater than 10 years of age in toy breeds.
Nutrition	Feeding of less than the recommended minimum percentages of protein (less than 22%) during pregnancy may be associated with smaller litter size as may deficiency in omega-6 and omega-3 fatty acids.
Inbreeding	Inbred bitches are more likely to carry small litters; this effect has been demonstrated in Dachshunds and Otterhounds.

Clinical implications

Medical records should be reviewed to determine cause of prior dystocia in any bitch to be bred so preventive measures can be taken if possible. Try to ensure optimal litter size by using serial measurement of serum progesterone to determine optimal breeding day; breeding bitches who are not excessively inbred, very young, or very old; ensuring appropriate nutrition throughout gestation; and carefully considering natural service versus artificial insemination (AI). If AI is used, carefully consider kind of semen and site of insemination.

Digital vaginal examination and pelvimetry may be useful components of the pre-breeding examination in trying to determine which bitches are predisposed to dystocia. Pelvimetry will only be a consistent and reliable technique when information about pelvic size in dogs of many breeds is reported.

Supplemental reading

Bergstrom A, Fransson B, Lagerstedt A-S, et al. 2006. Primary uterine inertia in 27 bitches: aetiology and treatment. *J Small Anim Pract* 47:456–460.

Bergstrom A, Nodtvedt A, Lagerstedt A-S, et al. 2006. Incidence and breed predilection for dystocia and risk factors for Cesarean section in a Swedish population of insured dogs. *Vet Surg* 35:786–791.

Johnston SD, Root Kustritz MV, Olson PNS. 2001. *Canine and Feline Theriogenology*. Philadelphia: WB Saunders Co., pp. 105–128.

Linde-Forsberg C. 2002. Pelvimetry to diagnose dystocia in the bitch. www.vin.com/proceedings/proceedings.plx?CID=WSAVA2002&PID=2684&Print=1, accessed August 23, 2008.

Root Kustritz MV. 2007. Decreased litter size in the bitch. Proceedings, Minnesota Veterinary Medical Association, Minneapolis, MN.

33

What are the normal physiologic and physical changes in bitches pre-partum?

General information

Estrogen concentrations are high during estrus, or heat, causing vulvar swelling and exudation of serosanguinous to straw-colored discharge, which originates in the uterus and drains through the vagina (see Chapter 20). Progesterone concentrations are high for the 2 months following estrus, regardless of pregnancy status. High serum progesterone is associated with closure of the cervix and lack of uterine contractility. Despite this, some normal bitches pass small volumes of serous or mucoid discharge from the vulva occasionally throughout gestation. Progesterone stimulates mammary development. As progesterone concentrations decline and prolactin concentrations rise very late in gestation, lactation begins. There is great variability in the ability to express milk from the mammary glands prior to parturition, with some bitches not having significant milk until after whelping.

Physiologic changes associated with pregnancy in bitches include normocytic, normochromic anemia, and hyperproteinemia. The normocytic, normochromic anemia of pregnancy is most likely due to an increase in plasma volume. Packed cell volume in late gestation varies from 24 to 35%; there may be a positive correlation between degree of anemia and number of fetuses. Plasma protein concentrations increase during pregnancy, with a decrease in total protein and albumin at the time of whelping. Other physiologic parameters reported to be altered in pregnant bitches are described in Table 33-1.

Glucose homeostasis is altered during pregnancy. Insulin resistance and decreased ability to synthesize glucose by gluconeogenesis and glycogenolysis lead to a relative increase in serum glucose concentrations and decreased ability to move glucose into cells. At its most extreme, this will be manifested as gestational diabetes. Subsequent lipolysis may manifest as ketosis and pregnancy toxemia (see Chapter 46).

Clinical implications

Passage of vulvar discharge during pregnancy can be normal. Microscopic examination of any discharge passed is strongly recommended to permit early diagnosis of pregnancy loss associated with inflammatory discharge (see Chapters 45 and 46). Bitches presenting as ill while pregnant should be worked up as any other dog, with knowledge of normal physiologic changes guiding interpretation of physical examination and blood-work results.

Table 33-1. Normal physiologic changes during pregnancy in bitches.

Physiologic Change	Increase/Decrease
Packed cell volume	Decreased—normocytic, normochromic
Plasma protein	Increased during gestation, decreased in late pregnancy
Albumin	Decreased in late pregnancy
Serum calcium	May be decreased in late pregnancy coincident with decreased albumin
Cardiac output	Increased
Blood volume/plasma volume	Increased
Hemoglobin concentration	Decreased
Oxygen consumption	Increased
Gastric emptying time	Increased
Glomerular filtation rate	Increased
Blood urea nitrogen and creatinine	Decreased

Supplemental reading

Johnson CA. 2008. Glucose homeostasis during canine pregnancy: Insulin resistance, ketosis and hypoglycemia. *Theriogenology* 70:1418–1423.

Johnston SD, Root Kustritz MV, Olson PNS. 2001. *Canine and Feline Theriogenology*. Philadelphia: WB Saunders Co., pp. 66–72.

Ryan SD, Wagner AE. 2006. Cesarean section in dogs: Physiology and perioperative considerations. *Compendium* 28:34–42.

34

How can I terminate pregnancy in bitches?

General information

When contacted by a bitch owner about mismating, one should first ascertain likelihood that the bitch was bred and became pregnant. Oftentimes, the breeding was not witnessed. If a breeding was witnessed, the bitch should be seen for vaginal cytology and drawing of blood for measurement of progesterone, to determine proximity to ovulation date (see Chapters 2 and 29). Lack of spermatozoa on vaginal cytology is not definitive proof that the bitch was not bred. Be aware that spermatozoa can live in the bitch's reproductive tract for up to 7 days, so if the bitch is identified to be in proestrus or estrus and periovulatory, she may become pregnant.

In surveys of bitches presented for pregnancy termination, 30 and 38% of the bitches were not pregnant. For this reason, and because recommended protocols work well at and after mid-gestation, pregnancy diagnosis is recommended before any therapy is instituted (see Chapter 5). For possible termination protocols, see Table 34-1. Drugs may be used in combination at the low end of the dose range to minimize side effects; for example, a prostaglandin may be combined with a prolactin inhibitor. No medical pregnancy termination protocol is reported to be 100% effective and no drugs are approved for pregnancy termination in bitches in the United States.

Clinical implications

Knowledge of stage of pregnancy best permits the veterinarian to tell the owner what to expect; for this reason, the author prefers ultrasound as a pregnancy diagnostic technique. If pregnancy is terminated before 40 days, virtually all evidence of pregnancy will be resorbed. From 40 to 50 days, fluid and tissue will be passed through the vulva. After 50 days, recognizable fetal tissues will be passed. After 55 days, pups may be born alive and should be euthanized upon passage. No one has demonstrated detriment to the bitch with pregnancy termination later in gestation but most veterinarians and owners will not perform termination late in pregnancy for aesthetic reasons.

When called with a question about mismating by an owner, the author uses the scheme outlined in Table 34-2.

Table 34-1. Pregnancy termination protocols reported for the bitch.

Drug	Regimen	Side Effects	Notes
Estrogens—suppress movement of eggs, affect implantation			
Estradiol cypionate ("mismate shot")	44 μg/kg intramuscular one time	1. Bone marrow toxicity—aplastic anemia, thrombocytopenia, leukopenia 2. Pyometra (see Chapter 47)	For treatment to be safe and effective, bitch must have ovulated (serum progesterone >4 to 10 ng/ml) and must still be in estrus (100% vaginal cornification). Bitches treated in diestrus may develop pyometra. There is no wisdom to treating with this drug within 3 days of mismating.
Prostaglandins—lyse corpora lutea (CL); decrease progesterone, cause uterine contractions			
Prostaglandin F2alpha (Lutalyse™)[a]	50 to 250 μg/kg subcutaneously twice to three times daily for 4 days or until pregnancy is terminated, identified by lack of fetal heartbeat or recognizable amniotic vesicles by ultrasound or passage of fetuses	1. Ptyalism, vomiting, diarrhea 2. Early return to estrus	Bitches will not respond to prostaglandin administered earlier than the 5th day after diestrus onset. Prostaglandin must be administered at least twice daily to effect lysis of the CL. Lower doses can be used later in gestation and are associated with decreased severity of side effects. Side effects usually occur and subside within 5–60 min of drug administration and lessen in severity with the series of injections.
Cloprostenol	2.5 μg/kg subcutaneously three times at 48-h intervals	As above	The author finds diarrhea to be more severe and less predictable in onset than with prostaglandin F2alpha.
Prolactin inhibitors—decrease function of CL (decrease progesterone)			
Bromocriptine	15 to 30 μg/kg twice daily per os after mid-gestation	Vomiting, anorexia	Bromocriptine is available in the United States as a 2.5 mg tablet (Parlodel™).
Cabergoline (Dostinex™)	5 μg/kg once daily per os after mid-gestation until pregnancy is terminated as described above	—	—

Table 34-1. *Continued*

Drug	Regimen	Side Effects	Notes
Progesterone receptor blockers			
Aglepristone (Alizine™)	10 mg/kg subcutaneously on 2 consecutive days after day 30	May cause early return to estrus	Not available in the United States.
Mifepristone	2.5 mg/kg twice daily per os after mid-gestation for 4.5 days	—	Available for human use only in the United States.
Corticosteroids			
Dexamethasone	0.1 to 0.2 mg/kg two to three times daily at a decreasing dose for 10 days after mid-gestation	1. Polyuria/polydipsia while receiving the medication 2. Tarry black vulvar discharge	Reported success rate is 80%. Pups that survive the protocol are reported to show no adverse effects from exposure to high levels of corticosteroids during development.

[a]The author has had good success with the following scheme: (1) diagnose pregnancy by ultrasound, verifying that bitch is at mid-gestation; (2) treat with prostaglandin F2alpha, 250 µg/kg subcutaneously twice daily for 4 days; (3) draw blood at time of last injection to verify progesterone has fallen to less than 1 ng/ml; and (4) repeat ultrasound in 1 week to verify pregnancy loss.

Table 34-2. Pregnancy termination management scheme.

1a. Breeding was witnessed. 2
1b. Breeding was not witnessed. 3
2a. The bitch was bred within the last 7 days. 3
2b. The bitch was bred more than 7 days ago. 4
3. The bitch should be seen for determination of likelihood of breeding having occurred. Vaginal cytology should be performed to determine stage of the estrous cycle (see Chapter 2). If the bitch is in proestrus or estrus, blood should be drawn to determine proximity to ovulation (see Chapter 29). If it is considered likely that she may have conceived. 4
4. Ovariohysterectomy can be performed at any time. If the owner does not desire the bitch to be sterilized, pregnancy diagnosis should be performed at mid-gestation (see Chapter 5). The only medical pregnancy termination therapy available that is effective around the time of conception and in very early pregnancy is estrogen; the author does not recommend use of this drug (see Table 34-1). The most commonly available medical therapy for pregnancy available in the United States at this time is prostaglandin F2alpha (Lutalyse™); please note that no drug is not approved for this use in bitches in the United States.

Supplemental reading

Corrada Y, Garcia P, De la Sota PE, et al. 2005. Decrease of body temperature after aglepristone treatment in bitches. *Anim Reprod Sci* 87:295–299.

Corrada Y, Rodriguez R, Tortora M, et al. 2006. A combination of oral cabergoline and double cloprostenol injections to produce third-quarter gestation termination in the bitch. *J Am Anim Hosp Assoc* 42:366–370.

Eilts BE. 2002. Pregnancy termination in the bitch and queen. *Clin Tech Small Anim Pract* 17:116–123.

Galac S, Kooistra HS, Butinar J, et al. 2000. Termination of mid-gestation pregnancy in bitches with aglepristone, a progesterone receptor antagonist. *Theriogenology* 53:941–950.

Gobello C, Castex G, Corrada Y, et al. 2002. Use of prostaglandins and bromocriptine mesylate for pregnancy termination in bitches. *J Am Vet Med Assoc* 220:1017–1019.

Johnston SD, Root Kustritz MV, Olson PNS. 2001. *Canine and Feline Theriogenology*. Philadelphia: WB Saunders Co., pp. 179–188.

Root Kustritz MV. 2006. *The Dog Breeder's Guide to Successful Breeding and Health Management*. St. Louis, MO: Elsevier, pp. 170–173.

Shafer-Somi S, Aksoy OA, Beceriklisoy HB, et al. 2007. Repeated induction of abortion in bitches and the effect on plasma concentrations of relaxin, progesterone and estradiol-17beta. *Theriogenology* 68:889–895.

35

Should bitches be spayed at the time of removal of a mammary mass?

General information

Mammary neoplasia is the most common tumor type in female dogs. Malignant tumors (adenocarcinoma) and benign tumors (fibroadenomas or mixed mammary tumors) occur with equal frequency. In humans, genetics and estrogen are causative factors. In bitches, no consistent breed predisposition has been reported. Similarly, not all canine mammary tumors contain estrogen receptors. However, early ovariohysterectomy (OHE) is associated with decreased incidence of mammary neoplasia in aged bitches. The protective effect of OHE is greatest if surgery is performed before the bitch goes through estrus, with a decreasing benefit with each heat cycle.

Surgery is the treatment of choice for mammary neoplasia. Concurrent OHE is often recommended. Arguments are that decreased exposure to estrogen and progesterone with subsequent heat cycles may prevent recurrence of neoplasia, and that performing both surgeries at once minimizes total number of anesthetic episodes in the bitch's life. Studies evaluating the beneficial effect of OHE at the time of surgical removal of mammary tumors have differed in results, with some showing increased survival time in bitches undergoing concurrent OHE and other studies showing no effect on survival time or biological behavior of the tumor.

Clinical implications

There is no overwhelming evidence to support concurrent OHE and mammary tumor removal as a method of decreasing mammary tumor recurrence and prolonging survival time. However, incidence of other conditions of aged bitches, specifically pyometra, is decreased by OHE. The greatest benefit may be minimizing the number of anesthetic episodes for a given bitch. If OHE and mammary tumor removal are performed at the same time, OHE should be performed first to prevent seeding of the abdomen with tumor cells.

Supplemental reading

Chang S-C, Chang C-C, Chang T-J, et al. 2005. Prognostic factors associated with survival two years after surgery in dogs with malignant tumors: 79 cases (1998–2002). *J Am Vet Med Assoc* 227:1625–1629.

Johnston SD, Root Kustritz MV, Olson PNS. 2001. *Canine and Feline Theriogenology*. Philadelphia: WB Saunders Co., pp. 246–253.

Osipov NE, Lagova ND, Ponomarkov VI. 1972. Spontaneous mammary gland tumors of dogs as a model for experimental tumor therapy. *Bull Exp Biol Med* 74:948–949.

Smith FO. 2006. Canine pyometra. *Theriogenology* 66:610–612.

Sorenmo KJ, Shofer FS, Goldschmidt MH, et al. 2000. Effect of spaying and timing of spaying on survival of dogs with mammary carcinoma. *J Vet Intern Med* 14:266–270.

36

Are there dietary supplements proven to improve semen quality in male dogs?

General information

Compounds commonly described as nutritional supplements that enhance semen quality in dogs are those containing glucosamine, most commonly in preparations for joint health, and those containing amino acids, specifically L-carnitine. While carnitine is secreted in the canine epididymis, the site of maturation of spermatozoa, the author is unaware of any published studies documenting supplementation with carnitine to enhance motility of spermatozoa or any other quality measure. Omega-3 fatty acid supplementation has been demonstrated to improve motility of frozen-thawed spermatozoa in stallions and to improve motility and sperm concentration in the boar. Similar studies have not been performed in dogs. Treatment of subfertile men with an antioxidant preparation has been associated with increased linear velocity of spermatozoa and increased pregnancy rate, compared with a placebo group. In rats, consumption of pomegranate juice, which contains vitamin C and several antioxidant compounds, was associated with improved semen quality. Vitamin, mineral, and herbal therapies marketed for enhancement of libido or semen quality have not been proven to be of benefit in dogs; however, no scientific studies have been reported.

Semen quality, specifically number of spermatozoa in the ejaculate, is improved by collection of semen in the presence of an estrous teaser bitch. In the absence of a teaser, the smooth muscle-contracting drug, prostaglandin F2alpha (Lutalyse™), given at a dose of 0.1 mg/kg subcutaneously 10 to 15 min before semen collection is attempted, has been demonstrated to increase number of spermatozoa in the ejaculate. Mild side effects may be observed, including salivation and panting. Similarly, gonadotropin releasing hormone may be given at a dose of 1 to 2 μg/kg subcutaneously 60 min before semen is collected; this presumably works by increasing serum testosterone concentrations and may be best used in apprehensive or shy males.

Clinical implications

Use of nutritional supplements for subfertile dogs is not supported by the veterinary literature. Veterinarians and their clients must weigh potential benefits against any known risks before using these compounds.

Supplemental reading

Comhaire FH, El Garem Y, Mahmoud A, et al. 2005. Combined conventional/antioxidant "Astaxanthin" treatment for male infertility: A double blind, randomized trial. *Asian J Androl* 7:257–262.

Hess M. 2006. Documented and anecdotal effects of certain pharmaceutical agents used to enhance semen quality in the dog. *Theriogenology* 66:613–617.

Turk G, Sonmez M, Aydin M, et al. 2008. Effects of pomegranate juice consumption on sperm quality, spermatogenic cell density, antioxidant activity and testosterone level in male rats. *Clin Nutr* 27:289–296.

37

Can I locate retained testes without surgery?

General information

Cryptorchidism is a common condition in dogs, with a reported incidence of 1.2 to 6.8%. One or both testes may be retained; in unilateral cryptorchids, the right testis is the one most commonly retained. Testes are more commonly retained in the inguinal area than in the abdomen. Retained testes are usually small and atrophic, and are more likely to undergo neoplastic change when the dog is aged than are descended testes.

The small size of retained testes makes them difficult to identify by palpation or diagnostic imaging. There are no reports of successful identification of non-neoplastic retained testes using radiography. Ultrasonography may be used, especially to confirm that a palpable inguinal mass is a retained testis and not a superficial inguinal lymph node. Although testicular tissue has a unique ultrasonographic appearance (see Chapter 8), abdominally retained testes are atrophic and may be difficult to identify unless torsion of the spermatic cord or neoplasia has caused testicular enlargement. Pathology of the testis may distort architecture of the tissue, causing a mixed echogenicity. Because location of retained testes is variable, ultrasonography is not commonly used to differentiate bilaterally cryptorchid dogs from castrated dogs.

Clinical implications

Differentiation of bilaterally cryptorchid dogs with abdominally retained testes from castrated dogs may be achieved by (1) palpation of the prostate—the prostate atrophies within weeks of surgery in castrated males and shows normal age-related hypertrophy in males with testes; or by (2) elevated serum testosterone concentrations with challenge testing—administer 1 to 2 µg/kg gonadotropin releasing hormone intramuscularly and draw blood for testosterone assay 60 min later. Any value greater than 3 ng/ml strongly suggests that at least one testis is present.

Although it may be possible to identify retained testicular tissue using ultrasound, surgical exploration and castration must be recommended to remove that male from the breeding pool. Decision tree analysis has demonstrated that the life span of cryptorchid dogs left intact does not vary significantly from dogs castrated at 1 year of age. However, unilateral cryptorchids are fertile, and this is a hereditary disorder. In one study evaluating genetic counseling for dog breeders, breeders voiced general disinterest in putting long-term breed or population concerns above short-term personal concerns regarding choice of stud dog in their kennel. It has also been demonstrated that choosing sires and dams to increase litter size may increase incidence of cryptorchidism in a given kennel. Veterinarians should stress the need for castration of all cryptorchid dogs, whether unilateral or bilateral.

Supplemental reading

Birchard SJ, Nappier M. 2008. Cryptorchidism. *Compendium* 30:325–336.

Davidson EB, Reaugh HF, Bahr RJ. 2006. What is your diagnosis: Cryptorchidism in a dog. *J Am Vet Med Assoc* 228:203–204.

Gubbels EDJ, Scholten J, Janss L, et al. 2008. Relationship of cryptorchidism with sex ratios and litter sizes in 12 dog breeds. *Anim Reprod Sci.* doi:10.1016/j.anireprosci.2008.07.004.

Johnston SD, Root Kustritz MV, Olson PNS. 2001. *Canine and Feline Theriogenology*. Philadelphia: WB Saunders Co., pp. 313–317.

Ortega-Pacheco A, Rodriguez-Buenfil JC, Segura-Correa JC, et al. 2006. Pathological conditions of the reproductive organs of male stray dogs in the tropics: Prevalence, risk factors, morphological findings and testosterone concentrations. *Reprod Dom Anim* 41:429–437.

Peters MAJ, Van Sluijs FJ. 2002. Decision analysis tree for deciding whether to remove an undescended testis from a young dog. *Vet Rec* 150:408–411.

Van Hagen MAE, Janss LLG, Van den Broek J, et al. 2004. The use of a genetic-counseling program by Dutch breeders for four hereditary health problems in boxer dogs. *Prev Vet Med* 63:39–50.

Yates D, Hayes G, Heffernan M, et al. 2003. Incidence of cryptorchidism in dogs and cats. *Vet Rec* 152:502–504.

38

Is there a medical therapy to cause testes to descend into the scrotum after 8 weeks of age?

General information

Normal testicular descent occurs in three phases. Abdominal translocation is downward migration and maintenance of the testis near the internal inguinal ring as the abdominal cavity elongates. The testis and epididymis on each side are attached to the gubernaculum, which mediates movement by growth and subsequent dehydration and shrinking of this gelatinous connective tissue. Physical presence of the testis is necessary at this stage, suggesting testicular weight, and secretion of testosterone and other secretory factors is important for this part of the process to occur. Transinguinal migration is movement of the epididymis and testis through the abdominal wall, and is mediated by intra-abdominal pressure. Because the inguinal canal does not close until 6 months of age in dogs, on average, transinguinal migration of testes is possible until the said age and dogs should not be definitely diagnosed as cryptorchid until they are older than 6 months. Inguinoscrotal migration is movement of the epididymis and testis into the scrotum. Testosterone is not necessary for this final stage, although it may play a role in the development of supportive structures guiding testicular movement.

Many factors play a role in testicular descent. Genetic factors most certainly play a role. In dogs, mode of heritability is undefined but testicular descent is most likely multi-genetic. The trait can be carried by bitches or dogs. Epigenetic factors (alterations in gene expression) and environmental factors also play a role.

Retained testes are significantly more likely to become neoplastic than are descended testes. Because of this predisposition to neoplasia and the heritable component of pathogenesis, bilateral castration is strongly recommended. Cryptorchid dogs cannot be shown in conformation dog shows sanctioned by the American Kennel Club. Administration of drugs or performance of surgery to cause testicular descent may permit showing of a dog with a heritable defect. The only ethically defensible argument for such treatment is to cause descent of the testes into the scrotum to allow castration through a pre-scrotal incision, rather than to require a more invasive abdominal exploration.

Compounds that have been reported to cause testicular descent in dogs include gonadotropin releasing hormone (50 to 750 μg one to six times in dogs aged 2 to 4 months, success rate 26.6% of 301 dogs), and human chorionic gonadotropin (100 to 1000 IU intramuscularly four times in a 2-week period in dogs less than 16 weeks of age, success rate 84% of 25 dogs, or 300 to 1000 IU three to four times in dogs of an unspecified age, success rate 75%).

Clinical implications

No single compound has been demonstrated consistently to cause testicular descent in dogs. Artificially inducing testicular descent is an ethical concern, as it may fraudulently increase the value of a dog with a heritable defect.

Supplemental reading

Amann RP, Veeramachaneni DNR. 2007. Cryptorchidism in common eutherian mammals. *Reproduction* 133:541–561.

Berkow R (ed.). 1982. *The Merck Manual*. Rahway, NJ: Merck and Co., Inc., p. 1609.

Feldman EC, Nelson RW. 1996. *Canine and Feline Endocrinology and Reproduction*. Philadelphia: WB Saunders Co., pp. 697–710.

Johnston SD, Root Kustritz MV, Olson PNS. 2001. *Canine and Feline Theriogenology*. Philadelphia: WB Saunders Co., pp. 313–317.

Section IV

Canine disease

39

How do I diagnose and treat dystocia?

Etiology

Dystocia, abnormal whelping, is best defined by comparison with normal whelping. Normal whelping occurs in three stages. Stage I of labor is the long stage during which the cervix dilates. The bitch acts restless, may show nesting behavior, pants, is often inappetent, and may vomit. She may be very clingy or very reclusive. Duration is 12 to 24 h. Stage II is active contractions and passage of the fetuses. A small volume of clear, tan, or slightly blood-tinged fluid may be passed before any pups are born. Once active contractions are seen, a pup should be born within 4 h; if the bitch is pushing hard constantly, a pup should be passed within 30 min. No more than 2 h should be allowed to pass between pups. Pups may be passed within the amniotic sac or the sac may rupture as the pup passes through the birth canal. At birth, the bitch should tear open the sac, if present, and vigorously lick the pup to stimulate respiration. Stage III is passage of the placentas. Stages II and III often alternate. Placentas should be passed within about 15 min of passage of a pup. The normal canine placenta is green in color due to pigmentation in the marginal hematomas on the edge of the placenta. The bitch may try to eat the placentas; this is an atavistic behavior that serves to keep the den clean to avoid attraction of predators (the placentas have no nutritional or therapeutic value).

Maternal causes of dystocia include uterine inertia, inadequate size of the birth canal, and abnormality of the pregnancy. Uterine inertia can be primary (no propulsive uterine contractions) or secondary (muscle fatigue after unproductive uterine contractions). Primary uterine inertia may be associated with hypocalcemia and may have a hereditary component. Low endogenous serum oxytocin concentrations may be a cause of primary uterine inertia in bitches with normal serum calcium concentration and may exacerbate the condition in bitches with low serum calcium concentrations. Secondary uterine inertia usually occurs in the presence of an obstruction preventing passage of a pup into the birth canal. Inadequate size of the birth canal may occur in bitches with previous pelvic fracture or a mass in the birth canal. The most commonly described abnormality of pregnancy associated with dystocia in bitches is uterine torsion.

Fetal causes of dystocia include fetal oversize and abnormalities of presentation, posture, and position. Fetal oversize may be absolute (normal birth canal, oversize pup) as in small litters or in bitches with gestational diabetes mellitus, or relative (part of pup too big to fit through normal birth canal) as in pups with hydrocephalus. Presentation is defined by which part of the pup first enters the birth canal. Cranial presentation (muzzle and extended forelimbs presenting to the birth canal) and caudal presentation (tail and extended rearlimbs presenting) are normal. Examples of abnormal presentations include breech (tail and flexed limbs presenting) and transverse (spine presenting). Position is defined by disposition of the extremities relative to the body. An example of abnormal position is neck flexed to the side.

Other disorders associated with dystocia are prolonged gestation (more than 64 days from ovulation or 71 days from a single breeding); drop in rectal temperature of greater than 1 F more than 24 h ago; and abnormal vulvar discharge (green discharge prior to the birth of any pups is indicative of placental separation).

Clinical signs

Bitches in labor may look tired but should not appear systemically ill. Disorientation, vomiting, or aggression are all abnormal during active labor in bitches. Normal vulvar discharge may be clear, tan, or slightly blood-tinged. It should not be purulent or frankly hemorrhagic, and should not be green if no pups have been passed.

Diagnosis

History findings suggestive of dystocia are prolonged time limits (longer than 24 h of stage I labor; more than 4 h since onset of stage II labor or more than 30 min since onset of active, repetitive contractions; more than 2 h between pups; more than 24 h since body temperature fell by full 1 °F, indicative of decline in serum progesterone concentrations), overtly abnormal presentation of a pup, or appearance of systemic illness or abnormal vulvar discharge from the bitch.

A brief but complete physical examination should be performed. A digital vaginal examination will not permit assessment of cervical dilation, as the canine vagina is too long for human fingers to reach the cervix. Circular constrictions palpated on digital vaginal examination are rings of vaginal constriction. A pup may be palpable in the birth canal.

Radiographs permit assessment of number and size of pups remaining in the uterus, and give some indication of presentation and posture (Fig. 39-1). Viability cannot be determined

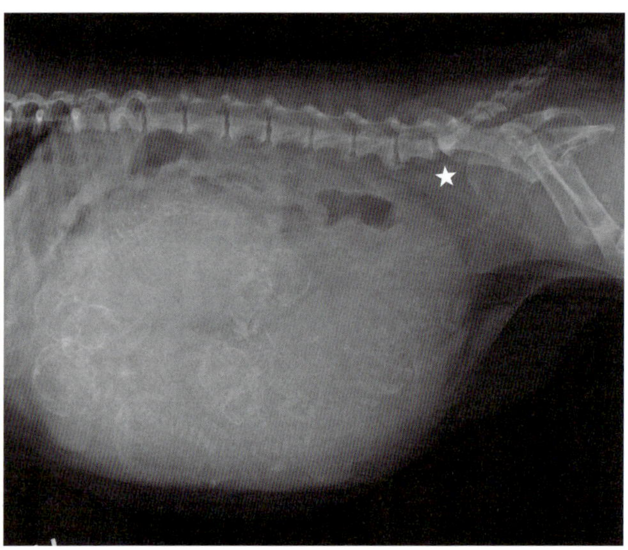

Figure 39-1: Pregnancy radiograph. Note head of pup in the birth canal (star).

accurately from radiography alone; signs of fetal death (gas within and around the fetus, collapse of the skull and axial skeleton) may not appear for up to 24 h after fetal death. Abdominal ultrasound is best used to determine viability by assessment of fetal heart rate. Normal fetal heart rate should be about twice that of the bitch. Fetal heart rate less than 170 bpm is indicative of fetal distress.

External parturition monitors are available for use in bitches. These devices measure changes in intra-abdominal pressure and permit interpretation of uterine contractility (see Resources).

Treatment

Manipulation of pups with abnormal presentation or position is difficult in dogs due to size restrictions. It is easy to dislocate joints and pull skin off pups while attempting manipulation. Instruments should not be used.

Oxytocin therapy is appropriate if radiographs have demonstrated normal size and likelihood that the pups can pass through the birth canal. Doses of 2 to 5 IU can be given intramuscularly at 20- to 30-min intervals for no more than three doses. Doses of greater than 5 IU cause uterine tetany, not propulsive uterine contractions. Repeated doses do not promote continuing uterine contractions because the uterus becomes refractory to oxytocin therapy as receptors fill on the myometrium and do not readily dissociate. Suggested initial doses are 0.25 IU for dogs weighing less than 11 lb (5 kg), 0.5 to 1.0 IU for dogs weighing 11 to 22 lb (5 to 10 kg), 1 to 3 IU for dogs weighing 22 to 66 lb (10 to 30 kg), and 3 to 5 IU for dogs weighing more than 66 lb (30 kg). Oxytocin increases frequency of uterine contractions; calcium increases strength of uterine contractions. Some bitches that do not respond to oxytocin therapy alone respond well to oxytocin and calcium given concurrently.

Table 39-1. Should the bitch be treated medically or surgically?

This scheme should be abandoned and Cesarean section performed if green vulvar discharge is evident prior to the birth of any pups, if fetal heart rate (by ultrasound) is less than 170 bpm, or if the bitch appears systemically ill.

Key for dystocia management:

1a. The puppy is present in the birth canal and can be manipulated for delivery 2
1b. The puppy is not present in the birth canal or cannot be manipulated for delivery 3
2a. Attempt delivery with lubrication and gentle traction. After that pup is passed or if other pups are present in utero 3
2b. Attempt delivery with lubrication and gentle traction. If the pup cannot be delivered 4
3a. Fetal heart rate is less than 150 bpm 4
3b. Fetal heart rate is 150 bpm or more 5
4. Perform Cesarean section.
5a. Abdominal radiographs have been taken 6
5b. Abdominal radiographs have not been taken 7
6a. Pups are too large to pass or are malpositioned 4
6b. Pups are not too large to pass and are not malpositioned 8
7. Take abdominal radiographs and go to 6
8a. Four or fewer pups are present 9
8b. More than four pups are present 4
9. Oxytocin therapy may be attempted as follows: Give 2 to 5 IU intramuscularly (IM); watch for effect for 20 min. If no effect is seen, give 2 to 5 IU oxytocin IM plus a 5-ml bolus of 10% calcium gluconate subcutaneously and watch for effect for 20 min. If no effect seen 4

Bitches with pups in distress, that are carrying pups too large to pass through the birth canal, or that are nonresponsive to oxytocin therapy should undergo Cesarean section (see Chapter 17 and 18; Table 39-1). In one study, 63.8% of bitches with dystocia underwent Cesarean section as therapy.

Supplemental reading

Bergstrom A, Frannson B, Lagerstedt A-S, et al. 2006. Primary uterine inertia in 27 bitches: Aetiology and treatment. *J Small Anim Pract* 47:456–460.

Bergstrom A, Nodtvedt A, Lagerstedt A-S, et al. 2006. Incidence and breed predilection for dystocia and risk factors for Cesarean section in a Swedish population of insured dogs. *Vet Surg* 35:786–791.

Davidson AP. Uterine and fetal monitoring in the bitch. 2001. *Vet Clin North Am* 31:305–313.

Johnston SD, Root Kustritz MV, Olson PNS. 2001. *Canine and Feline Theriogenology*. Philadelphia: WB Saunders Co., pp. 108–124.

Pretzer SD. 2008. Medical management of canine and feline dystocia. *Theriogenology* 70:332–336.

Root Kustritz MV. 2007. Use lower doses of oxytocin for dystocia. *Vet Med* 102:163.

40

How do I diagnose and treat eclampsia (hypocalcemia)?

Etiology

Eclampsia in dogs is more properly termed hypocalcemia, as it is a constellation of clinical signs associated with decreased calcium concentrations in serum and presumably at the cellular level. The condition in dogs should not be confused with pre-eclampsia in pregnant women, which is a disorder associated with high blood pressure and proteinuria.

Hypocalcemia develops as the fetuses mineralize and to an even greater extent when the bitch lactates. Lactation is a huge drain on body calcium stores. Bitches cannot take in enough calcium through the gastrointestinal tract to support milk production and so must draw calcium stores from the bone through the action of parathyroid hormone. Inadequate calcium in the extracellular compartment and subsequent decrease in membrane-bound calcium leads to spontaneous and repeated depolarization of muscle, or tetany.

Calcium supplementation during pregnancy may predispose bitches to hypocalcemia through down-regulation of parathyroid hormone. Diets balanced for calcium and phosphorus with ratios of 1:1 to 1.2:1 and that meet American Association of Feed Control Officials standards for gestation, lactation, and growth, or for all stages of life, are recommended during pregnancy.

Clinical signs

Clinical eclampsia is by far most common in small-breed bitches nursing large litters in their second or third week of lactation. However, any breed can be affected and clinical signs of hypocalcemia may occur during parturition (see Chapter 39). Inadequate maintenance of total body calcium is manifested initially by restlessness, muscle weakness or tremors, facial pruritus, and neglect of the pups. As the condition advances, tachycardia or bradycardia, and dilation of the pupils may be seen. End-stage eclampsia is characterized by convulsions and subsequent hyperthermia.

Diagnosis

Decreased total blood calcium (less than 7 mg/100 ml) or significantly decreased ionized calcium (less than 0.8 mmol/l) is diagnostic for hypocalcemia. Ionized calcium is the biologically active fraction of total calcium and its value cannot be extrapolated from direct or adjusted total calcium measurement. Hypomagnesemia and hyperkalemia may also be present.

If calcium measurement is not readily available in practice, treatment may be instituted based on signalment, history, and clinical signs. Body temperature should be evaluated in all dogs presenting with seizures.

Treatment

Hyperthermic animals (body temperature greater than 105°F) should be cooled slowly on presentation. Seizures should be controlled as for any cause, with administration of diazepam intravenously or other sedative or anesthetic agents.

Hypocalcemia is treated with intravenous calcium gluconate (10%) to effect. Most common doses described are 0.22 to 0.44 ml/kg or a bolus of 5 to 10 ml. The heart should be ausculted or an electrocardiogram strip monitored while calcium is administered intravenously, and therapy stopped as clinical signs resolve or if any abnormality of heart rate or rhythm is noted. Because calcium is irritating, caution should be taken to be aware of concentration of calcium administered.

After initial clinical signs have resolved, treatment with a depot of calcium intramuscularly or subcutaneously allows gradual absorption of calcium until oral calcium can be given. Oral calcium gluconate or calcium carbonate must be administered with vitamin D to ensure gastrointestinal absorption; doses are 1 to 3 g and 10,000 to 25,000 IU, respectively.

Oral therapy should continue until pups are weaned. Recurrence in the same lactation is not uncommon; if clinical eclampsia recurs in a given lactation, pups should be removed and hand-fed.

Supplemental reading

Aroch I, Srebro H, Shpigel NY. 1999. Serum electrolyte concentration in bitches with eclampsia. *Vet Rec* 145:318–320.

Drobatz KJ, Casey KK. 2000. Eclampsia in dogs: 31 cases (1995–1998). *J Am Vet Med Assoc* 217:216–219.

Johnston SD, Root Kustritz MV, Olson PNS. 2001. *Canine and Feline Theriogenology*. Philadelphia: WB Saunders Co., pp. 141–143.

Schenck PA, Chew DJ. 2005. Prediction of serum ionized calcium concentration by use of serum total calcium concentration in dogs. *Am J Vet Res* 66:1330–1336.

Schenck PA, Chew DJ. 2008. Hypocalcemia: A quick reference. *Vet Clin North Am* 38:455–458.

41

What are the causes of prolonged vulvar discharge after whelping in bitches?

Etiology

Normal postpartum discharge in bitches, termed lochia, is mucoid in character, varies in color from blood-tinged to green or brown, is odorless, and decreases in volume over time, with complete resolution by 3 weeks after whelping on average. Postpartum vulvar discharge may be considered abnormal due to character or duration of exudation.

Postpartum purulent discharge occurs most commonly as a manifestation of metritis. Metritis is a simple uterine infection, usually due to ascending infection with a microorganism from the vaginal flora after some insult to the uterus such as dystocia or retained feti or placentas.

Hemorrhagic discharge is associated with subinvolution of placental sites (SIPS), as is excessive duration of vulvar discharge after whelping. Subinvolution of placental sites is, as the name implies, a disorder associated with slow or abnormal healing of the endometrium after loss of the endotheliochorial placenta at parturition. Trophoblast-like cells in the uterus continue to invade the endometrium and prevent normal thrombus formation in endometrial blood vessels. Young bitches whelping their first litter are the ones most commonly diagnosed with SIPS; this may be because people choose not to breed affected bitches a second time.

Clinical signs

Metritis is characterized by exudation of purulent vulvar discharge, fever associated with inappetance and neglect of pups, and other clinical signs of systemic disease. The occasional bitch will present in hypovolemic or septic shock. SIPS is characterized by prolonged exudation of bloody vulvar discharge that is not inflammatory in nature when examined microscopically. Bitches with SIPS show no clinical signs of systemic disease.

Diagnosis

Metritis is diagnosed by cytology and culture of the vulvar discharge. Remember that the vagina is not sterile; a significant culture result is moderate to heavy growth of a single organism. Abdominal radiographs or ultrasound may permit determination of an underlying cause of ongoing inflammation, such as retained fetal material or fluid in the uterus. Complete blood count may reveal a neutrophilia. Hypoglycemia may be present in septicemic bitches.

SIPS is a rule-out diagnosis. Bitches with persistent vulvar discharge postpartum should be tested for brucellosis (see Chapter 44) and evaluated for metritis. Definitive diagnosis requires histopathology of the placental sites and so is usually reserved for bitches treated with ovariohysterectomy (OHE). Other rule-outs to consider when evaluating a bitch for persistent postpartum hemorrhagic vulvar discharge are coagulopathy (congenital, such as VonWillebrand's disease, or acquired, such as anticoagulant rodenticide toxicity), trauma, and vaginal neoplasia.

Treatment

Most bitches with metritis have a simple infection and recover well with antibiotic therapy guided by culture and sensitivity. Empiric therapy can be instituted with ampicillin (22 mg/kg per os twice daily) or amoxicillin-clavulanate (14 mg/kg per os twice daily). Bitches with severe systemic disease may require fluid replacement and intravenous antibiotic therapy. Evacuation of the uterus may be attempted if significant fluid or fetal remnants are present. Oxytocin rarely is effective at inducing uterine contractions beyond the first several days postpartum. Prostaglandin F2alpha (Lutalyse™) is effective at inducing uterine contractions at any time postpartum (50 to 250 μg/kg subcutaneously once daily to effect). Doses at the high end of the dose range are associated with more severe side effects. Side effects of prostaglandin F2alpha therapy in dogs are referable to contraction of the gastrointestinal tract and include hypersalivation, vomiting, and diarrhea. Signs appear within minutes of injection and subside within 60 min. Care should be taken when considering administration of any drug to cause uterine contractions, as the infected uterus may be friable. The very occasional ill bitch may become shocky when treated with prostaglandin F2alpha; it is strongly recommended that bitches be treated in the hospital and observed for a time after the drug is administered.

There is no specific medical therapy for SIPS. Most bitches do well with observation and periodic assessment of complete blood count for anemia and evaluation of the uterus to ensure that infection has not developed. Discharge resolves in most bitches before the subsequent proestrus. There are reports in the literature of the use of uterine contracting agents, such as oxytocin or ergonovine derivatives. These may cause pooling of blood within the uterus and predispose to infection and so are not recommended by this author. Therapy with progestogens has been reported with the idea that stimulation of the endometrium will hasten sloughing of residual trophoblast-like cells. Success with such therapy has not been consistent. Bitches with severe SIPS or that are not intended for further breeding should undergo OHE.

Supplemental reading

Johnston SD, Root Kustritz MV, Olson PNS. 2001. *Canine and Feline Theriogenology*. Philadelphia: WB Saunders Co., pp. 129–131, 139–141.

42

What is the best treatment for puppy vaginitis? Should the dog be allowed to go through one heat cycle?

Etiology

Puppy or juvenile vaginitis is most commonly an incidental finding during physical examination of prepuberal female dogs. The vulvar lips are glued together with a sticky, odorless, clear to white discharge. Puppy vaginitis may consist of inflammation due to urinary tract disease or abnormal vaginal anatomy with secondary infection. Often, puppy vaginitis is idiopathic.

Clinical signs

Most dogs present with small amount of mucoid, odorless vulvar discharge. Discharge may be caught at the vulvar lips or in the perivulvar hair. Occasionally, owners will describe seeing a string of discharge hanging from the vulva when the bitch urinates. Some bitches exhibit vulvar licking. Owners should be questioned about any evidence of urinary incontinence.

Diagnosis

Evaluation of the urinary tract and vagina should be performed concurrently to permit localization of the primary disease process, if possible. Diagnostic tests should include urinalysis and urine culture on samples collected by cystocentesis; digital vaginal examination and vaginoscopy, which may require sedation in dogs experiencing pain or dogs with a small vaginal vault; and microscopic examination of vulvar discharge. Diagnosis of vaginitis requires demonstration of inflammation in the vaginal vault. Normal vaginal mucosa should have the color of normal oral mucosa; inflamed mucosa varies from dark pink to dark red (see Chapter 3, Figs. 3-3 and 3-4). The vulvar discharge of dogs with juvenile vaginitis is mucoid with scattered inflammatory cells (Fig. 42-1).

Positive urine culture and inflammatory urine sediment are suggestive of underlying or concurrent urinary tract infection. Vaginal strictures or septa are often palpable on digital vaginal examination as a blockage prohibiting passage of a gloved finger into the vaginal vault. An underlying cause for vaginitis may not be identified in some cases.

Treatment

No treatment is necessary in dogs with minimal clinical signs. Twice daily cleaning of the perivulvar area with baby wipes or a nonalcohol-based otic cleanser may be beneficial. For severely affected dogs, treatment options are as follows.

- Appropriate antibiotic therapy, based on culture and sensitivity of a vaginal specimen. If empirical antibiotic therapy is required, oral amoxicillin-clavulanate has been demonstrated effective against 91 to 100% of bacteria commonly involved in vaginitis in female dogs.

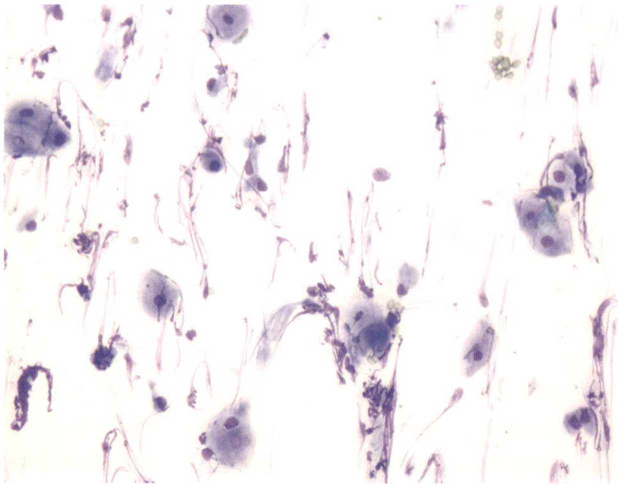

Figure 42-1: Vaginal cytology from a prepuberal dog with vaginitis.

- Correction of underlying causes of inflammation, if identified. Urinary tract infections should be treated with appropriate antibiotics, based on urine culture and sensitivity, and recurrent urinary tract infections investigated by radiography or cystoscopy to look for uroliths or other inciting causes of persistent disease. Vaginal anatomic anomalies should be defined by vaginography before surgical repair is attempted. Surgery is rarely performed in female dogs before they reach mature size.
- Treatment of possible subclinical urinary incontinence with secondary urine pooling and irritation of the vaginal mucosa. The drug used is phenylpropanolamine (1.0 to 1.5 mg/kg per os two to three times daily). Estrogen drugs, for example diethylstilbestrol, should not be used in prepuberal dogs as they may hasten physeal closure.
- Douching, or flushing of the vaginal vault, has not been demonstrated as an effective therapy for vaginitis in dogs.
- Dogs with clinical signs of atopy (pruritus, alopecia, licking at the paws, recurrent otitis) may benefit from therapy for atopy (diphenhydramine at 2 to 4 mg/kg per os three times daily, or hydroxyzine at 1 to 2 mg/kg per os two to three times daily; with either drug, taper dose if treatment is effective or if the side effect of sedation is severe).

There is no science to guide the answer regarding suitability of permitting a female dog to go through one heat cycle before ovariohysterectomy (OHE) as a possible cure for puppy vaginitis. Anecdotal reports suggest that some bitches with persistent prepuberal vaginitis resolve their disease after going through one heat cycle. Whether this is due to the hormonal changes of estrus altering the vaginal epithelium or just to maturity of the dog and her immune function is not known. This decision must be made on a case-by-case basis, remembering that the primary benefit of prepuberal OHE is a substantial decrease in predisposition to mammary neoplasia when aged (see Chapter 22).

Supplemental reading

Johnston SD, Root Kustritz MV, Olson PNS. 2001. *Canine and Feline Theriogenology*. Philadelphia: WB Saunders Co., p. 235.

Root Kustritz MV. 2008. Vaginitis in dogs: A simple approach to a complex condition. *Vet Med* 103:562–567.

43

What is the best treatment for chronic vaginitis in a spayed female dog?

Etiology

Infection with one or more species of the normal bacterial flora overlies a primary cause of inflammation. The most common underlying causes of inflammation in dogs with vaginitis are urinary tract infection and vaginal anomalies.

Clinical signs

Vaginitis in spayed female dogs is usually evidenced by mucoid to mucopurulent vulvar discharge. Discharge is rarely sanguinous; bloody vulvar discharge is more often associated with estrus or ovarian remnant syndrome, coagulopathy, trauma, neoplasia of the genitourinary tract, or blood parasites. Spayed female dogs affected with vaginitis may demonstrate concurrent vulvar licking and urinary incontinence. Clitoral hypertrophy may develop secondary to vulvar licking (Fig. 43-1).

Diagnosis

Evaluation of the urinary tract and vagina should be performed concurrently to permit localization of the primary disease process, if possible. Diagnostic tests should include urinalysis and urine culture on samples collected by cystocentesis; digital vaginal examination and vaginoscopy, which may require sedation in dogs experiencing pain or dogs with a small vaginal vault; and microscopic examination of vulvar discharge. Vaginal cytology in affected dogs reveals non-cornified epithelial cells, mucoid debris, and polymorphonuclear cells (Fig. 43-2). Diagnosis of vaginitis requires demonstration of inflammation in the vaginal vault. Normal vaginal mucosa should have the color of normal oral mucosa; inflamed mucosa varies from dark pink to dark red (see Chapter 3, Figs. 3-3 and 3-4).

Positive urine culture and inflammatory urine sediment are suggestive of underlying or concurrent urinary tract infection. Vaginal strictures or septa are often palpable on digital vaginal examination as a blockage prohibiting passage of a gloved finger into the vaginal vault. An underlying cause for vaginitis may not be identified in some cases.

Differentials for mucopurulent vulvar discharge in bitches include onset of diestrus (see Chapter 2), *Brucella canis* infection (see Chapter 44), metritis (see Chapter 41), and pyometra (see Chapter 47).

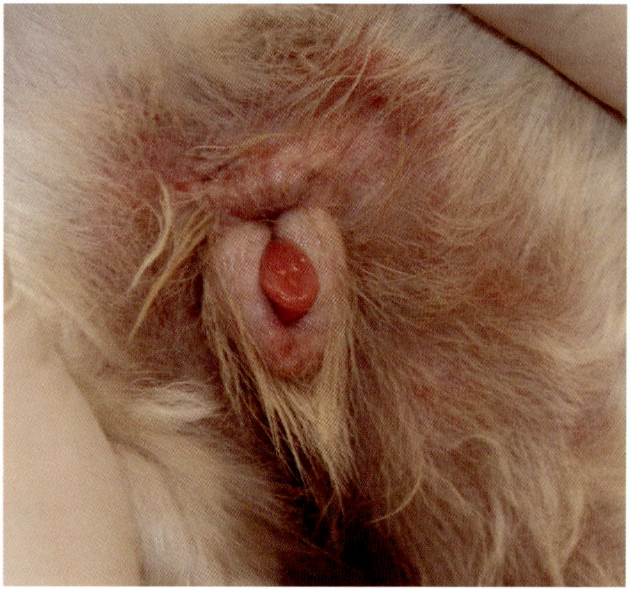

Figure 43-1: Clitoral hypertrophy.

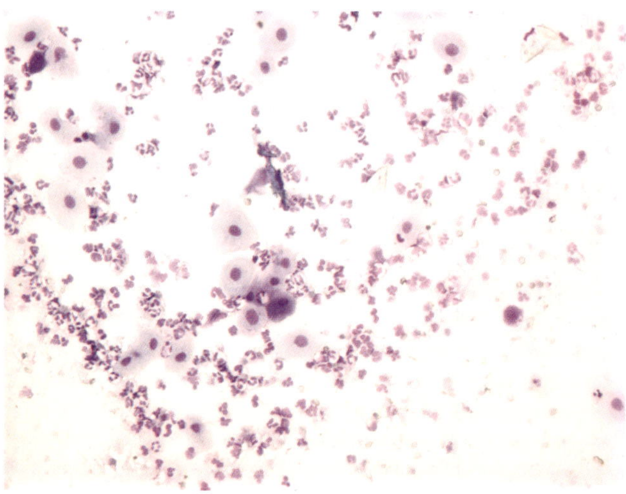

Figure 43-2: Vaginal cytology from an adult spayed dog with vaginitis.

Treatment

Treatment consists of:

- Appropriate antibiotic therapy, based on culture and sensitivity of a vaginal specimen

If empirical antibiotic therapy is required, oral amoxicillin-clavulanate has been demonstrated effective against 91 to 100% of bacteria commonly involved in vaginitis in female dogs.

- Correction of underlying causes of inflammation, if identified

Urinary tract infections should be treated with appropriate antibiotics, based on urine culture and sensitivity, and recurrent urinary tract infections investigated by radiography or cystoscopy to look for uroliths or other inciting causes of persistent disease. Vaginal anatomic anomalies should be defined by vaginography before surgical repair is attempted.

- Treatment of possible subclinical urinary incontinence with secondary urine pooling and irritation of the vaginal mucosa

Drugs used include diethylstilbestrol (0.1 to 0.2 mg/kg per os once daily for 5 days [maximum dose = 1 mg], tapering to every 4 to 7 days) or phenylpropanolamine (1.0 to 1.5 mg/kg per os two to three times daily).

- Glucocorticoid therapy may be attempted in bitches with no history of urinary incontinence. Specific dose regimens have not been established.
- Douching, or flushing of the vaginal vault, has not been demonstrated as an effective therapy for vaginitis in adult spayed female dogs.
- Dogs with clinical signs of atopy (pruritus, alopecia, licking at the paws, recurrent otitis) may benefit from therapy for atopy (diphenhydramine at 2 to 4 mg/kg per os three times daily, or hydroxyzine at 1 to 2 mg/kg per os two to three times daily; with either drug, taper dose if treatment is effective or if the side effect of sedation is severe).

Supplemental reading

Crawford JT, Adams WM. 2002. Influence of vestibulovaginal stenosis, pelvic bladder, and recessed vulva on response to treatment for clinical signs of lower urinary tract disease in dogs: 38 cases (1990–1999). *J Am Vet Med Assoc* 221:995.

Johnston SD, Root Kustritz MV, Olson PNS. 2001. *Canine and Feline Theriogenology*. Philadelphia: WB Saunders Co., p. 235.

Root Kustritz MV. 2008. Theriogenology question of the month: Diestrual discharge in a bitch. *J Am Vet Med Assoc* 232:841–842.

Root Kustritz MV. 2008. Vaginitis in dogs: A simple approach to a complex condition. *Vet Med* 103:562–567.

Wang KY, Samii VF, Chew DJ, et al. 2006. Vestibular, vaginal, and urethral relations in spayed dogs with and without lower urinary tract signs. *J Vet Intern Med* 20:1065–1073.

44

How do I test for and treat canine brucellosis?

Etiology

Canine brucellosis is most often due to infection with *Brucella canis*. Infection with other *Brucella* sp. may occur but is uncommon. Transmission is by ingestion or sexual contact, with organisms shed in all body fluids. Canine brucellosis is a reemerging disease, perhaps due to increased movement of dogs between states and countries.

Clinical signs

Many bitches and dogs carrying brucellosis are asymptomatic or present for infertility. Classical clinical signs in infected bitches are late-term pregnancy loss, birth of stillborn pups, and persistent purulent vulvar discharge. Classical clinical signs in male dogs are epididymitis and orchitis, with subsequent poor semen quality. Infection of other tissues may occur and be evidenced as uveitis, meningitis, or diskospondylitis.

Diagnosis

Culture

Culture of the organism is the testing modality that identifies disease soonest after infection, and is definitive. However, culture is difficult as contaminants often overgrow Brucella and there is a risk of infection of laboratory workers since brucellosis is a zoonotic disease.

Serologic testing

Serologic testing is performed more commonly than culture. Serologic tests may identify antibodies raised against Brucella antigens or may identify portions of the Brucella organism itself. Because of variations in specificity and sensitivity of available tests, more than one test may be required for definitive diagnosis (Table 44-1).

Agglutination tests

Agglutination tests identify antibodies raised against cell surface proteins of Brucella. Since Brucella shares cell surface proteins with many common bacteria, including Staph, Strep, and Bordetella, false positive results are common with this type of test. Agglutination tests do not become positive until about 8 to 12 weeks after infection. The rapid slide or rapid card agglutination test is easy to run in-house using instructions included in the kit (see Resources; Fig. 44-1).

CANINE DISEASE

Table 44-1. Diagnostic testing sequence for canine brucellosis.

Diagnostic Test	Negative Test Result	Positive Test Result	Interpretation
Rapid slide agglutination testing (RSAT)	X	—	The dog is not actively infected with *Brucella canis*.
	—	X	RSAT results often yield false positives. Testing with tests other than an agglutination test should be performed.
Agarose gel immunodiffusion (AGID) test	X	—	The dog is not actively infected with *B. canis*.
	—	X	AGID test results for cytoplasmic antigens are specific and positive results indicate active infection.

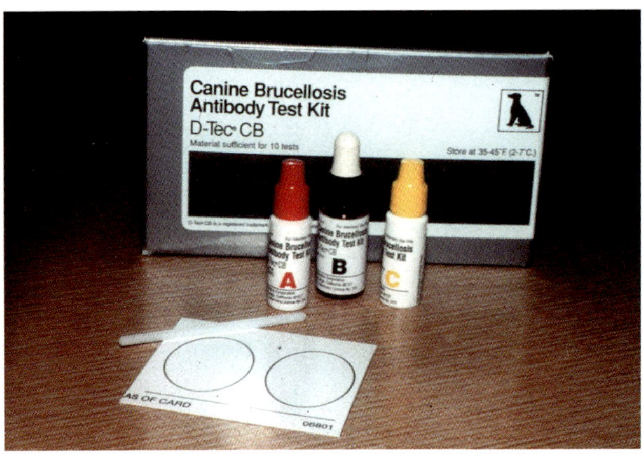

Figure 44-1: Rapid slide agglutination test for canine brucellosis (reprinted with permission from Root Kustritz MV. 2006. *The Dog Breeder's Guide to Successful Breeding and Health Management*. St. Louis, MO: Elsevier).

The tube agglutination test used at commercial laboratories provides a titer instead of a yes–no answer. Serologic titers of less than 1:100 are considered negative, of 1:100 to 1:200 suspects, and greater than 1:200 indicative of active infection. Agglutination tests are good screening tests. All negative results can be considered true negatives. All positive results should be reevaluated using a different kind of test.

Other testing modalities

Agarose gel immunodiffusion (AGID) testing is considered the gold standard by the Centers for Disease Control and Prevention. This technique tests for antibodies against both cell wall antigens, which are shared with other bacterial species, and cytoplasmic antigens, which are unique to *Brucella* sp. This test becomes positive about 12 weeks after infection and remains positive even after the animal becomes abacteremic, making it valuable for its accuracy and its ability to identify chronic infection. Positive test results are considered true positives. The AGID test for canine brucellosis is available at Cornell University and at the University of Georgia (see Resources).

Enzyme-linked immunosorbent assay (ELISA) tests are described in the literature, again identifying antibodies against either cell wall or cytoplasmic antigens. While sensitivity and specificity of these tests are very good, no ELISA tests are commercially available as of this writing.

Polymerase chain reaction (PCR) testing identifies minute amounts of Brucella proteins. It is a remarkably specific and very sensitive test. Current research has evaluated only naturally infected dogs so time from infection until the test becomes positive is not reported. As of this writing, there are no commercially available PCR tests for canine brucellosis in the United States.

Treatment

Infected animals are intermittently bacteremic and *B. canis* organisms may be sequestered inside cells. Most therapies described use a combination of a tetracycline and a streptomycin. Because streptomycin is not readily available, some protocols recommend use of a combination of a tetracycline and an aminoglycoside; all dogs to be subjected to this regimen must first be checked for normal renal function. Therapy with enrofloxacin as a sole agent has also been described. Any antibiotic therapy will cause a decrease in bacteremia and subsequent decrease in antibody titers. However, recrudescence of disease is common and infected dogs must be regularly tested and retreated throughout their lives. Because the disease is difficult, if not impossible, to eradicate

Table 44-2. Management of *Brucella canis* positive animals.

Dogs housed in kennels	The kennel should be closed, with no dogs in or out. All male and female dogs housed in the kennel, including dogs not actively used for breeding, should be screened with the rapid slide agglutination testing. All positives should be rechecked with the agarose gel immunodiffusion (AGID). All negatives should be housed away from any suspect or positive animals. Once a dog has tested negative for 3 months in a row, it should be housed in a specific area for negative animals. All true positives, identified by AGID testing, should be euthanized. Once all remaining animals in the kennel have tested negative for 3 consecutive months, the kennel can be opened.
Dogs housed in homes	Individual positive dogs should be spayed or castrated and housed away from all other dogs on the premises, or provided a home with no other dogs. The infected dogs should not have access to young children, elderly people, or immunocompromised people. Testing should be performed every 3 to 6 months and antibiotic therapy used as necessary to minimize bacterial shedding.

from an individual dog, spreads within kenneled dogs, and has zoonotic potential, euthanasia of positive animals should be considered, especially in kennel situations or in homes where the positive animal would be housed with other dogs or exposed to people more likely to become infected (young children, aged adults, immunocompromised children or adults; Table 44-2).

Supplemental reading

Allen A, Ngeleka M, Parker K. 2008. Brucellosis in kennels in western Canada. *Can Vet J* 49:625 [letter].

Hollett RB. 2006. Canine brucellosis: Outbreaks and compliance. *Theriogenology* 66:575–587.

Keid LB, Soares RM, Vasconcellos SA, et al. 2007. A polymerase chain reaction for detection of Brucella canis in semen of naturally infected dogs. *Theriogenology* 67:1203–1210.

Keid LB, Soares RM, Vasconcellos SA, et al. 2007. A polymerase chain reaction for detection of Brucella canis in vaginal swabs of naturally infected bitches. *Theriogenology* 68:1260–1270.

Keid LB, Soares RM, Vasconcellos SA, et al. 2009. Comparison of agar gel immunodiffusion test, rapid slide agglutination test, microbiological culture and PCR for the diagnosis of canine brucellosis. *Res Vet Sci* 86:22–26.

Wanke MM. 2004. Canine brucellosis. *Anim Reprod Sci* 82-83:195–207.

45

What are the causes of vulvar discharge during pregnancy?

Etiology

Vulvar discharge during pregnancy may arise from the urinary tract or the reproductive tract. Urinary tract conditions that may occur during pregnancy include cystitis and urolithiasis and, rarely, neoplasia. Reproductive tract conditions associated with vulvar discharge during pregnancy include brucellosis (see Chapter 44), pregnancy loss (see Chapter 46) and, rarely, pyometra (see Chapter 47). Vaginitis is not commonly described to occur during pregnancy in bitches and neoplasia of the genitourinary tract is most common in bitches beyond breeding age.

Clinical signs

Character of vulvar discharge guides diagnostics and treatment. A small volume of clear, odorless discharge may be normal, especially late in pregnancy. Urinary tract disease is usually associated with dysuria and hematuria. Bitches infected with *Brucella canis* may present with no other signs; classically, infected bitches abort after 45 days of gestation. Pregnancy loss may be associated with passage of dark, red to brown, odorless discharge or passage of visible fetal tissues. Pyometra is associated with passage of purulent discharge that may be blood-tinged.

Diagnosis

Urinary tract disease is diagnosed as in any animal; urine samples should not be collected by cystocentesis without ultrasound guidance to ensure that the uterus is not penetrated. Diagnostic tests include urinalysis and urine culture, radiography, and possibly urethroscopy and cystoscopy.

Serologic testing for brucellosis is recommended on any bitch exhibiting vulvar discharge during pregnancy (see Chapter 44). Causes of pregnancy loss and appropriate diagnostics are discussed in Chapter 46; drawing of blood for measurement of progesterone in serum and performance of a complete blood count (CBC), and collection of a sample of vulvar discharge for aerobic culture are minimum requirements. Pyometra is characterized by presence of purulent vulvar discharge, uterine enlargement defined by ultrasound, and signs of systemic disease including leukocytosis. Pyometra and pregnancy can occur concurrently.

Treatment

Treatment is dependent on cause. If all diagnostic tests are normal (negative serologic test for canine brucellosis, normal serum progesterone concentration, normal CBC, nonsignificant aerobic culture results) and the pups appear viable on ultrasound (fetal heart rate greater than 170 bpm), observation is recommended. Empirical antibiotic therapy will prevent ascension of vaginal flora through the open cervix and subsequent secondary infection. For treatment of brucellosis and other causes of pregnancy loss, and for pyometra, see the appropriate chapters.

Supplemental reading

Johnston SD, Root Kustritz MV, Olson PNS. 2001. *Canine and Feline Theriogenology*. Philadelphia: WB Saunders Co., pp. 225–228.

46

What diagnostic tests can be performed to identify the cause of pregnancy loss?

Etiology

Pregnancy loss can occur due to infectious or noninfectious causes. Infectious causes include bacterial infection with *Brucella canis* or other bacterial species, canine herpesvirus infection and other viral diseases, and protozoal infection with *Toxoplasma gondii* and perhaps other protozoal species. Noninfectious causes include hypoluteoidism, gestational diabetes mellitus, pregnancy toxemia, developmental defects, and uterine disease.

Clinical signs

Pregnancy loss is associated with vulvar discharge and may or may not be associated with signs of systemic disease. Oftentimes, pregnancy loss is the presenting complaint. Signs that may be seen with specific infections are described under Diagnosis.

Diagnosis

When presented with a pregnant dog that appears to be losing the pregnancy based on presence of vulvar discharge or systemic signs of disease, a complete physical examination, abdominal ultrasound, drawing of blood for measurement of concentration of progesterone in serum and complete blood count, and collection of a sample of vulvar discharge for aerobic culture should be performed. If fetal heart rate is less than 200 bpm, fetal distress is present and the fetuses should be closely monitored for viability. If fetal heart rate consistently falls to less than 150 to 170 bpm, fetal death is imminent and pregnancy loss should be allowed to continue or the pups removed via hysterotomy or hysterectomy.

If fetal tissues or aborted pups are available for diagnosis, tissues should be harvested as soon after death as possible and the tissues not frozen prior to examination. Fresh samples should be refrigerated and submitted to a diagnostic laboratory and fixed samples should be submitted as well. Tissues that should be submitted include liver, kidney, adrenal, small and large intestine, lung, heart, thymus, brain, fetal membranes, and fetal stomach content or fetal fluids. Swabs from the bitch's vagina should be submitted in transport media.

Infectious causes

Bacteria

B. canis is classically associated with late-term abortion and persistent purulent vulvar discharge. Testing is described in Chapter 44.

Other bacteria reportedly associated with pregnancy loss in bitches are *Salmonella* sp., *Campylobacter* sp., *Streptococcus* sp., *Listeria monocytogenes*, and leptospires. Clinical signs that may be present in bitches with salmonellosis include diarrhea, vomiting, fever, and depression. Dogs with leptospirosis may present with concurrent abortion, meningitis, and uveitis. Many bitches whose pregnancy loss is caused by bacterial infections present with vulvar discharge as the only clinical sign of disease. Samples to be collected for bacterial culture include vulvar discharge; passed fetal tissues; and stomach contents of aborted pups, which contain amniotic fluid and therefore accurately reflect the intrauterine environment.

Viruses

The most common viral infection associated with pregnancy loss in bitches is canine herpesvirus (CHV). Nonpregnant adult dogs exposed to the virus demonstrate few clinical signs, with small vesicles on mucosal surfaces and slight vulvar and nasal discharge. Exposed bitches develop enough memory T cells to permit them to respond immunologically to subsequent challenges. Bitches considered to be at greatest risk are those who were first exposed to the virus during the last 3 weeks of gestation. In these bitches, placentitis may develop and the bitch will lose pups or give birth to a mixed litter of mummified, macerated, stillborn, and apparently normal pups. Serologic testing is available but no commercially available test has excellent specificity and sensitivity. Because CHV is not a very antigenic virus, active infection must be identified by assessment of acute and convalescent titers. Carriers cannot be identified by measurement of antibody titers. Polymerase chain reaction tests, which identify portions of the CHV proteins, have proven disappointing as a screening or diagnostic tool. Definitive diagnosis requires demonstration of the pathognomonic signs of vasculitis in abdominal organs or virus isolation from infected pups (see Chapter 96, Fig. 96-1).

Protozoa

T. gondii is the most common protozoal infection associated with pregnancy loss in dogs. Systemic signs of disease are always present and include depression, diarrhea, ocular and nasal discharge, and anorexia. Definitive diagnosis requires demonstration of cysts of the organism in fetal tissues.

Other protozoa reportedly associated with pregnancy loss in bitches are *Neospora caninum* and *Leishmania* sp. The dog is the definitive host of *N. caninum* and reported seroprevalence is as high as 20% in some surveys. It appears that most dogs carry the organism without being clinically affected.

Noninfectious causes

Hypoluteoidism

Progesterone is required throughout pregnancy in all species. In bitches, all progesterone secretion arises from the corpora lutea (CLs) on the ovaries. Premature lysis of the CLs is associated with decreased serum progesterone and pregnancy loss. Progesterone concentrations rise immediately after ovulation, peak at mid-gestation, and gradually fall through the last month of pregnancy. If serum progesterone falls to less than 1 to 2 ng/ml for more than 24 h, pregnancy will not be maintained. If on initial assay serum progesterone is between 5 and 10 ng/ml, the author will retest in 1 to 2 days; if at any point serum progesterone falls to less than 5 ng/ml, supplementation with progesterone can be instituted if pups are still viable, as validated by abdominal ultrasound. There is no value in supplementing progesterone early in gestation or if the pups are not viable. In early pregnancy, progesterone may masculinize female fetuses and in the presence of nonviable fetuses, pregnancy loss should be allowed to commence, regardless of cause.

Gestational diabetes mellitus

Pregnancy and associated elevated serum progesterone concentrations are associated with insulin resistance and decreased ability of bitches to generate glucose from gluconeogenesis and glycogenolysis, predisposing them to diabetes mellitus. Diagnosis requires demonstration of hyperglycemia and glucosuria.

Pregnancy toxemia

Bitches late in pregnancy carrying large litters that become inappetent will rely on lipolysis to meet energy needs. Breakdown products of fat metabolism, including ketones, are released into the circulation and contribute to anorexia, worsening the condition. Diagnosis requires demonstration of ketonuria.

Developmental abnormalities

Older bitches, with aged ova, may be more likely to have small litters due to DNA abnormalities reflected as developmental defects noncompatible with life. Teratogenic drugs may also induce abnormalities that are incompatible with life; the latter is more commonly associated with loss of an entire litter while the former is more commonly reflected as loss of individual pups.

Uterine disease

Cystic endometrial hyperplasia (CEH) and chronic uterine infection may be associated with pregnancy loss. In bitches with severe uterine changes, chemical abnormalities may alter the function of spermatozoa. CEH may alter uterine motility and interfere with movement of the fertilized ova, and implantation or placentation of the embryos. Severe uterine abnormalities may be visible on ultrasound (see Chapter 47, Fig. 47-2). Definitive diagnosis requires uterine biopsy.

Treatment

Bacterial infections with organisms other than *B. canis* are treated with appropriate antibiotics, based on culture and sensitivity, and acknowledging danger of some drugs during pregnancy if live pups are present (see Chapter 30). *B. canis* is not readily cleared even with antibiotic therapy and euthanasia may be the best solution, especially in kennel situations (see Chapter 44).

Canine herpesvirus is not treatable in adult animals and there is limited information about treatment in infected pups (see Chapter 96). A vaccine is available in Europe; it is reported to improve fertility in endemically infected kennels.

Treatment for protozoal infection with *T. gondii* is usually not effective at preventing pregnancy loss.

Hypoluteoidism is treated with supplementation with progesterone. Injectable progesterone in oil can be assayed in serum to ensure adequate serum concentrations are being reached. However, daily injections may be necessary and since it is a depot preparation, it is difficult to assess speed with which the compound is absorbed and distributed. Oral forms of progesterone available include synthetic and natural compounds. Synthetic compounds cannot be assayed in serum but are readily absorbed and excreted, permitting greater control of administration. Natural compounds can be assayed and are readily absorbed and excreted. Micronized progesterone (Prometrium; see Resources) is administered at a dose of 10 mg/kg two to four times daily per os until day 60 to 62 after the luteinizing hormone surge (day 58 to 60 after ovulation), or until the day of planned Cesarean section. Serum progesterone concentrations should be assessed every 2 to 3 days during therapy to ensure adequate serum concentrations are being maintained.

In bitches affected with gestational diabetes mellitus, prognosis for the pregnancy is poor, with increased puppy mortality. Glycemic control with insulin therapy is difficult to achieve; pregnancy termination may be preferred. In one survey of 13 cases of gestational diabetes mellitus in bitches, diabetes mellitus persisted after completion of pregnancy in four (30.8%).

Pregnancy toxemia therapy requires movement of the bitch from a negative to a positive energy balance. This may be affected by force-feeding or parenteral nutrition. Most commonly, pregnancy is terminated to permit the bitch to normalize metabolism without the burden of the fetuses.

Supplemental reading

Fall T, Kreuger SJ, Juberget A, et al. 2008. Gestational diabetes mellitus in 13 dogs. *J Vet Intern Med* 22:1296–1300.

Givens MD, Marley MSD. 2008. Infectious causes of embryonic and fetal mortality. *Theriogenology* 70:270–285.

Johnson CA. 2008. Glucose homeostasis during canine pregnancy: Insulin resistance, ketosis, and hypoglycemia. *Theriogenology* 70:1418–1423.

Johnson CA. 2008. High-risk pregnancy and hypoluteoidism in the bitch. *Theriogenology* 1424–1430.

Pretzer SD. 2008. Bacterial and protozoal causes of pregnancy loss in the bitch and queen. *Theriogenology* 70:320–326.

Root Kustritz MV. 2005. Pregnancy diagnosis and abnormalities of pregnancy in the dog. *Theriogenology* 64:755–765.

Rube A, Corrada Y, Arias D, et al. 2007. Progesterone insufficiency in bitches in early and late diestrous. *Vet Rec* 160:524–525.

Schlafer DH. 2008. Canine and feline abortion diagnostics. *Theriogenology* 70:327–331.

Verstegen J, Dhaliwal G, Verstegen-Onclin K. 2008. Canine and feline pregnancy loss due to viral and non-infectious causes: A review. *Theriogenology* 70:304–319.

47

How do I diagnose and treat pyometra in bitches?

Etiology

Pyometra is a two-step process. The first pathologic change is cystic endometrial hyperplasia (CEH), a thickening of the uterine lining that occurs secondary to repeated estrous cycling. The unique estrous cycle of the bitch, with high serum estrogen followed by prolonged elevation in progesterone after every cycle, promotes hyperreactivity of the endometrium and gradual cystic hyperplasia. By 9 years of age, two-thirds of bitches have discernible CEH. Development of CEH is a continuum and it is not completely understood why some bitches develop pyometra with minimal CEH while other bitches with severe CEH do not develop pyometra.

The second pathologic change is infection. Infection invariably is due to an organism that is part of the normal vaginal flora. *Escherichia coli* is the most common isolate. Studies have demonstrated that it is the bitch's own subtype of *E. coli* that is associated with infection; she does not become infected because of exposure to a male. Pyometra is most common in bitches that have never been pregnant, suggesting that pregnancy may have some sort of protective effect, presumably at the level of the endometrium.

The most likely sequence is as follows: (1) the bitch cycles repeatedly, with increasing CEH over time; (2) when the bitch goes through proestrus and estrus, bacteria ascend from the vagina into the uterus; (3) the altered endometrium does not readily permit expulsion of all organisms before the end of estrus; (4) the bitch goes into diestrus, during which the cervix closes, trapping bacteria; secretion of endometrial glands is increased, nourishing those bacteria; and the endometrium becomes hyperreactive, exacerbating the hyperplasia; and (5) infection develops with creation of a pool of purulent intrauterine fluid. It is not known what factors determine patency of the cervix.

Renal disease is a common sequel to pyometra. Endotoxins released from the cell wall of Gram-negative bacteria inhibit normal renal tubular function. These changes are reversible if infection is controlled quickly enough. Another reported disorder that may be associated with pyometra is peritonitis; in one study, 6 of 11 intact females with secondary peritonitis had pyometra as the primary disease condition.

Clinical signs

Pyometra is most common in older, nulliparous bitches who have been in heat within the last 4 to 12 weeks. Clinical signs vary with cervical patency.

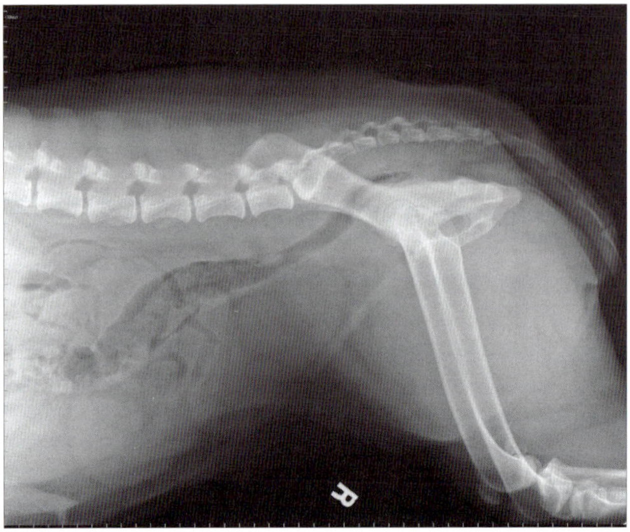

Figure 47-1: Lateral abdominal radiograph from bitch with pyometra.

Bitches presenting with open-cervix pyometra have yellow–green to pink or red-tinged, thick, odoriferous vulvar discharge. Other signs include depression, inappetence, and vomiting. Affected dogs may be febrile or may be hypothermic. Dogs with secondary renal disease will be polyuric with subsequent polydipsia (PU/PD).

Bitches presenting with closed-cervix pyometra do not have vulvar discharge. Systemic signs of disease are as with open-cervix pyometra (depression, anorexia, vomiting, PU/PD, hyper- or hypothermia) but usually are more severe. Abdominal distension is evident.

Diagnosis

Diagnosis of pyometra requires demonstration of excessive fluid in the uterus and either purulent vulvar discharge (open-cervix) or systemic response indicative of infection (closed-cervix). Uterine enlargement can be identified by abdominal palpation; care should be taken in manipulating the friable uterus. Imaging is less directly traumatic to the uterus. Radiography can be used to identify uterine enlargement but cannot differentiate disease from pregnancy in animals who were bred (Fig. 47-1). Ultrasonography is preferred because it permits the operator to see within the uterus to identify free fluid (Fig. 47-2). Pregnancy is readily identifiable using ultrasound by 24 to 25 days after ovulation, permitting definitive differentiation from disease. Ultrasound cannot be used to determine if intrauterine fluid is purulent.

Cytology of vulvar exudate reveals full fields of degenerative polymorphonuclear cells (PMNs), bacteria, and non-cornified epithelial cells (Fig. 47-3). A sample of this discharge should be submitted for aerobic culture and sensitivity.

Blood should be drawn for a complete blood count and serum chemistry profile. Approximately 77% of dogs present with a leukocytosis with a left shift. The number of PMNs present may be very high, especially in dogs with closed-cervix pyometra. Changes that may be evident on serum chemistry profile include azotemia and hyperproteinemia.

Urine should not be collected by cystocentesis without ultrasound guidance. Free-catch urine may be contaminated with vulvar discharge. Changes evident on urinalysis include decreased urine specific gravity and proteinuria.

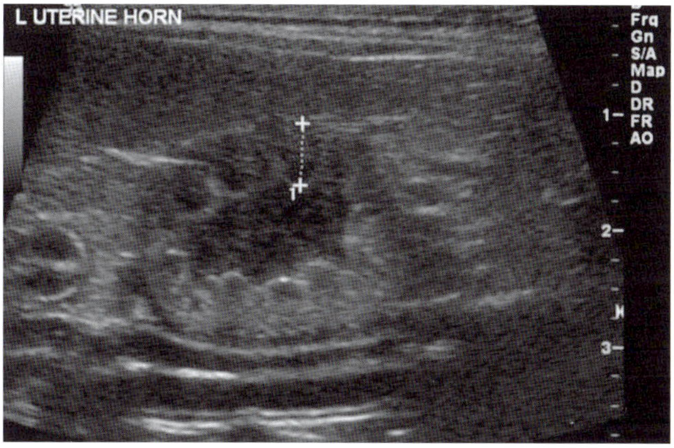

Figure 47-2: Ultrasonogram from bitch with pyometra. Line delineates thickened uterine wall.

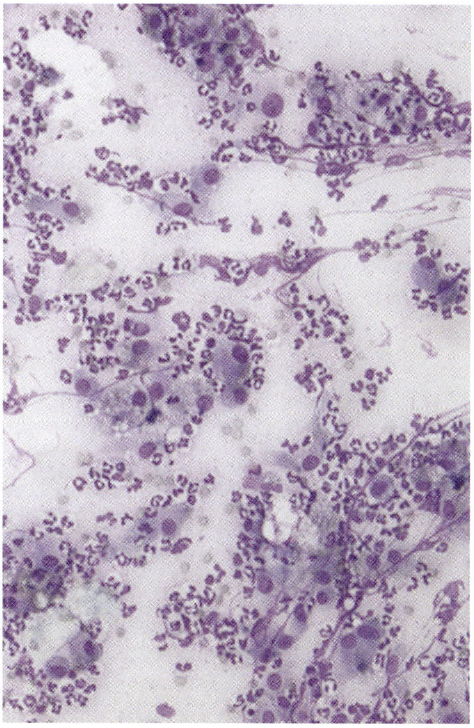

Figure 47-3: Cytology of vulvar discharge from a bitch with pyometra.

Treatment

Ovariohysterectomy (OHE) is the best treatment for pyometra in all cases. Surgical removal of the infected uterus immediately clears endotoxins from the body, with return to normal white count and lymphocyte activity by 7 days after surgery. CEH is not reversible so no medical therapy can return the dog to normal reproductive function. Extreme proteinuria after OHE

Table 47-1. Medical treatment of canine pyometra.

1. Assess uterine size in a repeatable manner (palpation, radiography, ultrasonography).
2. Collect a specimen of vulvar discharge for aerobic culture and sensitivity testing. Begin empirical therapy with ampicillin or amoxicillin. Change antibiotic if necessary based on culture and sensitivity results.
3. Draw blood for measurement of progesterone in serum. If serum progesterone is less than 2 ng/ml, once daily therapy with prostaglandin will be effective. If serum progesterone is greater than 2 ng/ml, twice daily therapy with prostaglandin will be required to lyse the corpora lutea and decrease serum progesterone concentration, permitting more effective uterine contractions.
4. Administer prostaglandin F2alpha (0.1 to 0.25 mg/kg subcutaneously once or twice daily). Treat until uterine size nears normal (palpation, radiographs) or until no free intrauterine fluid is visible (ultrasound).
5. Send the bitch home on antibiotics. Antibiotic therapy should continue for 1 month or until there has been no visible vulvar discharge for at least 1 week.
6. The bitch is predisposed to pyometra at every subsequent cycle. Early in the next proestrus, a sample should be collected from the cranial vagina with a guarded swab and antibiotic therapy instituted. The bitch should be bred at that first cycle after medical therapy for pyometra and should be spayed as soon as her breeding life is over. Bitches that cannot be bred on a given cycle should be treated with an appropriate antibiotic as long as the cervix is open, evidenced by vulvar discharge.

(urine protein:creatinine ratio greater than 10) is associated with increased likelihood of end-stage renal failure in the dog's future.

Medical therapy may be considered if the bitch meets the following criteria: (1) the cervix is open; (2) azotemia is absent or mild enough to be attributed to dehydration or other pre-renal causes; (3) the bitch is still of breeding age (generally less than 6 years of age); and (4) the bitch is a valuable part of a breeding program. Medical therapy is not recommended in cases of closed-cervix pyometra. Treatment with drugs that cause or permit uterine contractions may help push purulent fluid against the internal cervical os with subsequent relaxation but these drugs may also promote movement of fluid through the uterine tubes and into the abdomen, or cause uterine rupture, with subsequent peritonitis.

Medical therapy with antibiotics and prostaglandin F2alpha is described here (Table 47-1). Combined therapy using a synthetic prostaglandin (cloprostenol; 5 µg/kg every third day subcutaneously [SQ]), prolactin inhibitor (cabergoline; 5 µg/kg once daily per os), and antibiotics was described as successful in treating either open- or closed-cervix pyometra in 19 of 22 bitches. Other possible therapies include use of a long, rigid endoscope to pass a polypropylene urinary catheter through the cervix and promote drainage; use of the drug aglepristone (Alazine™), a progesterone receptor blocker (10 mg/kg SQ days 1,2, 8, and 14 with concurrent antibiotic therapy); acupuncture; and herbal therapy. These therapies are not well described in the veterinary literature or are not available in the United States.

Supplemental reading

Bartoskova A, Vitasek R, Leva L, et al. 2007. Hysterectomy leads to fast improvement of haemotological and immunological parameters in bitches with pyometra. *J Small Anim Pract* 48:564–568.

Culp WTN, Zeldis TE, Reese MS, et al. 2009. Primary bacterial peritonitis in dogs and cats: 24 cases (1990–2006). *J Am Vet Med Assoc* 234:906–913.

Egenvall A, Hagman R, Bonnett BN, et al. 2001. Breed risk of pyometra in insured dogs in Sweden. *J Vet Intern Med* 15:530–538.

England GCW, Freeman SL, Russo M. 2007. Treatment of spontaneous pyometra in 22 bitches with a combination of cabergoline and cloprostenol. *Vet Rec* 160:293–296.

Fieni F. 2006. Clinical evaluation of the use of aglepristone, with or without cloprostenol, to treat cystic endometrial hyperplasia-pyometra complex in bitches. *Theriogenology* 66:1550–1556.

Gurbulak K, Pancarci M, Ekici H, et al. 2005. Use of aglepristone and aglepristone + intrauterine antibiotic for the treatment of pyometra in bitches. *Acta Vet Hung* 53:249–255.

Hagman R, Kindahl H, Lagerstedt A-S. 2006. Pyometra in bitches induces elevated plasma endotoxin and prostaglandin F2alpha metabolite levels. *Acta Vet Scand* 47:55–68.

Hagman R, Kuhn I. 2002. Escherichia coli strains isolated from the uterus and urinary bladder of bitches suffering from pyometra: Comparison by restriction enzyme digestion and pulsed-field gel electrophoresis. *Vet Microbiol* 84:143–153.

Heiene R, Kristiansen V, Teige J, et al. 2007. Renal histomorphology in dogs with pyometra and control dogs, and long-term clinical outcome with respect to signs of kidney disease. *Acta Vet Scand* 49:13–21.

Hildebrandt TB, Goritz F, Boardman W, et al. 2006. A non-surgical uterine lavage technique in large cats intended for treatment of uterine infection-induced infertility. *Theriogenology* 66:1783–1786.

Johnston SD, Root Kustritz MV, Olson PNS. 2001. *Canine and Feline Theriogenology*. Philadelphia: WB Saunders Co., pp. 206–224.

Jurka P, Max A, Hawrynska K, et al. 2008. Age-related pregnancy results and further examination of bitches after aglepristone treatment of pyometra. *Reprod Dom Anim* doi:10.1111/j.1439-0531.2008.01288.x.

Nam T-C, Kim H-Y, Seo K-M, et al. 2009. Non-surgical treatment for uterine pyometra using acupuncture in dogs. www.mondialvet99.com/WSAVA/cc/oraux/acupuncture/tchi_tchou.htm, accessed May 9, 2000.

Niskanen M, Thrusfield MV. 1998. Associations between age, parity, hormonal therapy and breed, and pyometra in Finnish dogs. *Vet Rec* 143:493–498.

Salgado B, Paramo R, Sumano H. 2007. Successful treatment of canine open cervix pyometra with yunnan-pai-yao, a Chinese herbal preparation. *Vet Res Commun* 31:405–412.

Schlafer DH, Gifford AT. 2008. Cystic endometrial hyperplasia, pseudo-placental endometrial hyperplasia, and other cystic conditions of the canine and feline uterus. *Theriogenology* 70:349–358.

Verstegen J, Dhaliwal G, Verstegen-Onclin K. 2008. Mucometra, cystic endometrial hyperplasia, and pyometra in the bitch: Advances in treatment and assessment of future reproductive success. *Theriogenology* 70:364–374.

48

What are the causes of agalactia in bitches and how is it best treated?

Etiology

Agalactia is lack of milk production or secretion. Reported causes of lack of milk production include inadequate nutrition, stress, progestogen therapy and subsequent inadequate prolactin secretion, endotoxemia, and systemic illness. The nutritional needs of lactating bitches are extreme, with reported increases in total energy intake during peak lactation 3.7 times that of pre-partum in Labrador retrievers and 2.6 times that of pre-partum in miniature Schnauzers.

Lack of milk letdown is due to inadequate oxytocin secretion. Treatment with repeated low doses of oxytocin intramuscularly or with human oxytocin nasal preparations often immediately results in expression of milk from the multiple teat openings on each gland in the bitch.

Clinical signs

Agalactia is evidenced by inability to express milk from the mammary glands, even after oxytocin injection has stimulated milk let-down.

Diagnosis

Diagnosis is by inspection.

Treatment

If an underlying cause can be identified, it should be treated. However, a cause such as malnutrition in the bitch often cannot be remedied in a timely enough manner to permit milk production. Treatment with the dopamine antagonist, domperidone, may stimulate prolactin secretion and lactation. This is not an approved use of this drug in any species. Acupuncture has been described as an effective therapy for agalactia in several species, including dogs.

Supplemental reading

Johnston SD, Root Kustritz MV, Olson PNS. 2001. *Canine and Feline Theriogenology*. Philadelphia: WB Saunders Co., pp. 138–139.

Root Kustritz MV. 2006. *The Dog Breeder's Guide to Successful Breeding and Health Management*. St. Louis, MO: Elsevier, pp. 230–231.

Scantlebury M, Butterwick R, Speakman JR. 2000. Energetics of lactation in domestic dog (Canis familiaris) breeds of two sizes. *Comp Biochem Phys Part A* 125:19–210.

49

How do I treat mastitis in a nursing bitch?

Etiology

Mastitis is most commonly seen postpartum in the actively nursing bitch. Occasional bitches with galactorrhea associated with false pregnancy may develop mastitis. Environmental organisms and skin flora are the bacteria most commonly associated with the disease and include *Escherichia coli*, *Staphylococcus* sp., and *Streptococcus* sp.

Clinical signs

The infected gland is engorged, reddened, hot, and painful to the bitch upon manipulation (Fig. 49-1). More than one gland may be affected. Fluid expressed from the multiple teat openings varies from apparently normal milk to purulent to clear or blood-tinged serous fluid (Fig. 49-2). The bitch may be febrile with inappetence and lethargy, and may neglect the pups. Localized abscesses may develop with gangrene of tissue (Fig. 49-3). Occasionally, the bitch will undergo septic shock.

Diagnosis

Mastitis is diagnosed by inspection. Cytology of fluid expressed from the mammary gland reveals bacteria and inflammatory cells. Expressed fluid from the affected gland(s) should be submitted for culture and sensitivity. Gangrenous tissue is black.

Treatment

Antibiotics likely to achieve therapeutic levels in the mammary gland are those that are lipid-soluble and are therefore also likely to be excreted into milk. Caution must be taken if using antibiotics with potential negative side effects on neonates (see Chapter 91). Pups should be allowed to nurse unless gangrenous mastitis is present; organisms implicated in septicemia in pups are rarely those cultured from milk. The affected gland(s) should be covered to prevent excoriation of the friable tissue by puppy nails. Common empirical choices are cephalexin (2.5 to 7.5 mg/lb [5 to 15 mg/kg] per os three times daily) and amoxicillin-clavulanate (Clavamox; 7 mg/lb [14 mg/kg] per os two to three times daily). Abscessed areas should be incised, gangrenous tissue debrided, and drains placed.

Pain management in lactating bitches is complicated by the ready movement of most analgesics into milk and the lack of research documenting potential effects on nursing puppies.

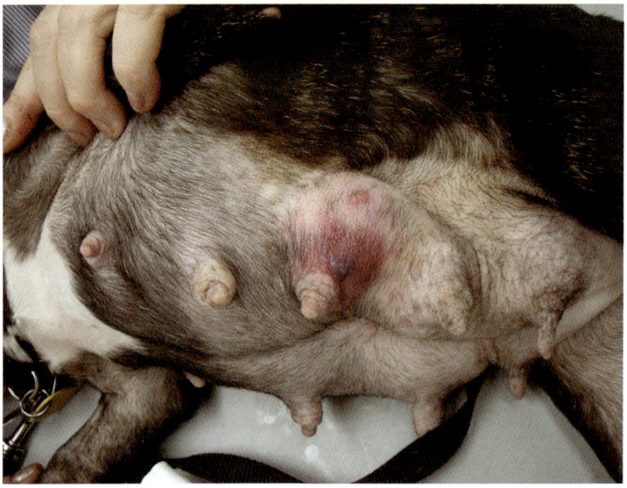

Figure 49-1: Mastitis in a lactating bitch.

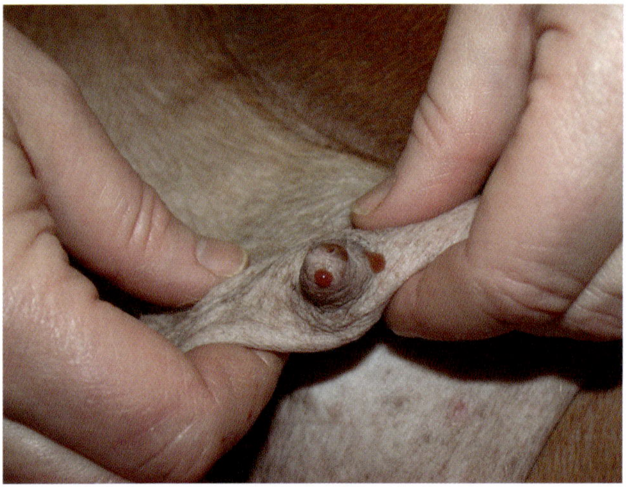

Figure 49-2: Abnormal fluid expressed from the mammary gland of a bitch with mastitis (photo courtesy of Dr. Cathy Gartley).

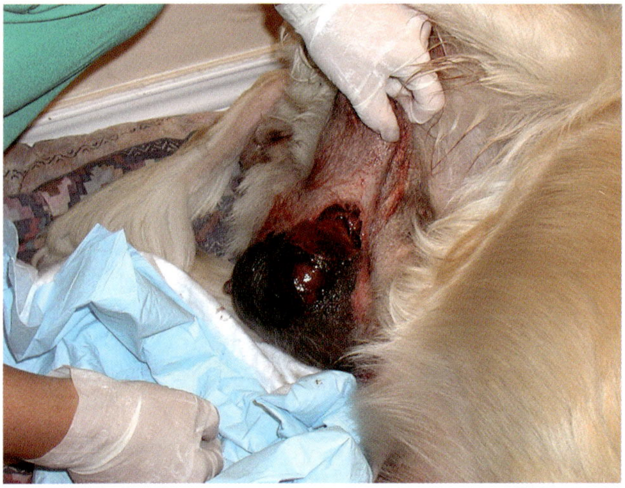

Figure 49-3: Gangrenous mastitis in a bitch (photo courtesy of Dr. Cathy Gartley).

Nonsteroidal anti-inflammatory analgesics may exacerbate postpartum hemorrhage or may inhibit renal maturation in neonates. Those with a long half-life (naproxen, piroxicam, miloxicam) may accumulate in neonates. Opioids also pass readily into milk but can be reversed in neonates if signs of overdosing become apparent (lethargy, failure to nurse).

Supplemental reading

Johnston SD, Root Kustritz MV, Olson PNS. 2001. *Canine and Feline Theriogenology*. Philadelphia: WB Saunders Co., pp. 131–138.

Mathews KA. 2008. Pain management for the pregnant, lactating, and neonatal to pediatric cat and dog. *Vet Clin North Am* 38:1291–1308.

Shafer-Somi S, Spergser J, Breitenfellner J, et al. 2003. Bacteriological status of canine milk and septicemia in neonatal puppies—A retrospective study. *J Vet Med B* 50:343–346.

50

What are the causes of persistent estrus (heat) in dogs?

Etiology

The most common causes of true persistent estrus in bitches are follicular ovarian cysts and ovarian tumors. Persistent estrus is caused by continuing secretion of estrogen. Estrogen production through the adrenal gland is possible but has not been reported as a pathologic phenomenon in bitches. Exogenous estrogen administration, either as a "mismating shot" (see Chapter 34) or through dermal absorption of topical products applied to the owner, is another possible cause.

Clinical signs

Remember that proestrus can last as long as 17 days and estrus as long as 21 days. This means that a normal bitch could exhibit signs of estrus for 4 to 6 weeks. True persistent estrus is defined as physical and behavioral signs of proestrus and estrus, and presence of cornified vaginal cytology, for at least 6 weeks (see Chapters 2 and 20). Bilaterally symmetrical alopecia may be evident, especially if exposure to estrogen has been very prolonged. Abdominal distension and ascites may be evident in bitches with granulosa cell tumor.

Diagnosis

Follicular ovarian cysts most commonly develop at the expected time of estrus. The cysts are visible by abdominal ultrasonography and are usually not large enough to distort the ovary (Fig. 50-1). The most common ovarian tumor type that is functional, secreting estrogen, is granulosa cell tumor. Clinical signs may arise at any time, not necessarily at the expected time of estrus. Oftentimes, the ovary is so enlarged as to be palpable per abdomen as a cranial abdominal mass. Ultrasonographically, ovarian tumors are more likely to be appear mottled and septate. Differentiation of follicular ovarian cysts and ovarian tumors may also be made by response to treatment. Similarly, diagnosis of exogenous estrogen as a source may be made by careful history taking and response to withdrawal of potential inciting agents. With any prolonged estrogen exposure, bone marrow suppression may occur, evidenced by non-regenerative anemia, leukopenia, and thrombocytopenia.

Treatment

Ovarian follicular cysts are treated by induction of ovulation, either with gonadotropin releasing hormone ([Cystorelin] 50 μg/dog intramuscularly [IM]) or human chorionic gonadotropin

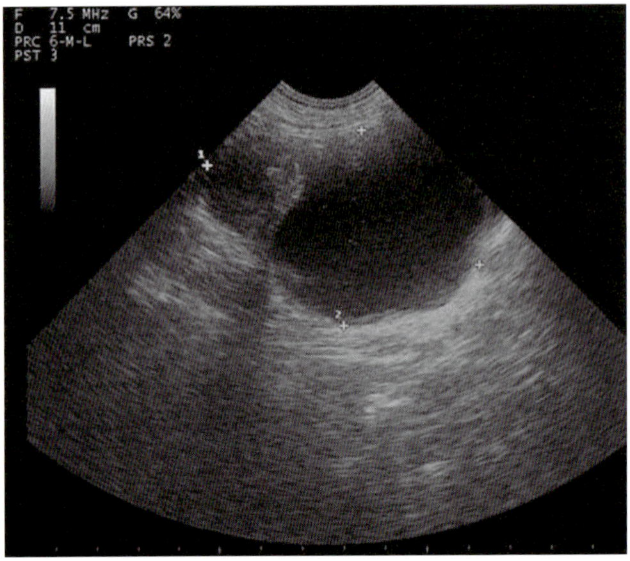

Figure 50-1: Ultrasonogram of a follicular cyst in a bitch.

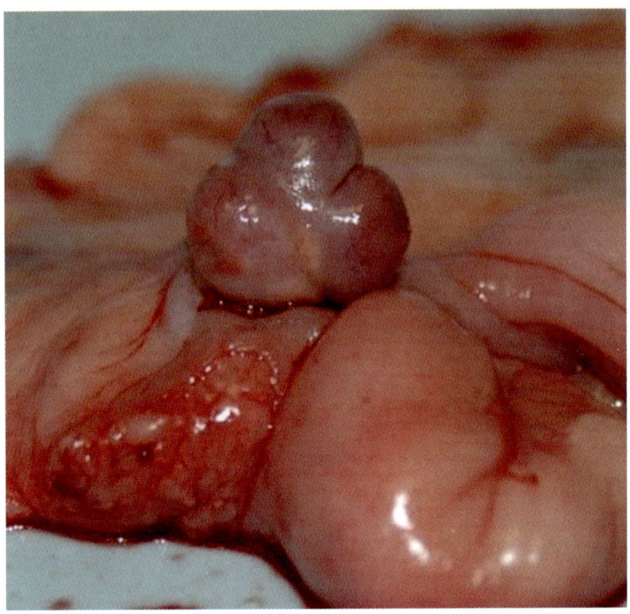

Figure 50-2: Gross appearance of a follicular cyst in a bitch.

(1000 IU, half intravenously, half IM or 22 IU/kg IM once daily for 3 days), or by ovariohysterectomy (OHE) (Fig. 50-2). OHE may be the preferred therapy as dogs induced to ovulate after persistent estrogen exposure may be predisposed to pyometra (see Chapter 47). Granulosa cell tumors are treated by OHE. Because metastasis and local invasion are uncommon, OHE is usually curative. Blood dyscrasias are treated with appropriate blood products.

Supplemental reading

Johnston SD, Root Kustritz MV, Olson PNS. 2001. *Canine and Feline Theriogenology*. Philadelphia: WB Saunders Co., p. 193.

Meyers-Wallen VN. 2007. Unusual and abnormal canine estrous cycles. *Theriogenology* 68:1205–1210.

Schwarze RA, Threlfall WR. 2008. Theriogenology question of the month: Exogenous estrogen exposure from a topically applied human product. *J Am Vet Med Assoc* 233:235–237.

51

What are the causes of lack of cycling (persistent anestrus) in dogs?

Etiology

Two forms of anestrus are described in bitches. Secondary anestrus is lack of estrous cycling within 1 year of a previous cycle. Causes include silent heat, systemic disease, possibly including hypothyroidism (see Chapter 27), poor body condition or excessive athleticism, administration of estrus-suppressing drugs, and luteal ovarian cysts. Primary anestrus is lack of any apparent estrous cycling in bitches 2 years of age or older. Causes are as for secondary anestrus and may also include chromosomal abnormalities and previous ovariohysterectomy (OHE).

Clinical signs

Regular observation of bitches is recommended, especially for those housed outside or in a separate facility. Periodic exposure to an intact male dog may aid the owner in identifying bitches in heat. Silent heat is, by definition, not apparent clinically. The ovaries go through the normal sequence of follicular development and ovulation but no overt physical or behavioral changes are observed. Animals with reproductive dysfunction associated with hypothyroidism rarely have classic clinical signs of hypothyroidism, such as bilaterally symmetrical alopecia and lethargy. Poor body condition or tremendous conditioning due to athleticism is obvious on physical examination.

Diagnosis

The owner should be questioned regarding where the bitch is housed and how he or she has tried to identify signs of heat. Other history questions may include diet and work schedule of the bitch, medications given, and if there is any possibility of the bitch having been spayed. Previous OHE may be identified by demonstration of a scar on the ventral midline or by demonstration of elevated concentrations of luteinizing hormone in the blood. Systemic disease may be identified by serum chemistry profile and complete blood count. Hypothyroidism is best diagnosed by demonstration of decreased free thyroxine and increased thyroid stimulating hormone (see Chapter 27). Blood should be drawn for assay of progesterone in serum; progesterone concentration of greater than 2 ng/ml may indicate either that a silent heat occurred within the last 2 months (the bitch is in diestrus) or that a luteal ovarian cyst is present. Luteal cysts are usually not visible ultrasonographically; persistence of elevation in serum progesterone for more than 2 months is diagnostic. Chromosomal abnormalities can be identified by karyotyping, which is available at Texas A&M University (see Supplemental Reading).

Treatment

If there is no clear evidence of systemic disease or another cause of lack of cycling, bitches can be put on a good plane of nutrition, retired for a time from showing or working, and housed with other cycling bitches, who may pull them into heat through pheromone exposure (the dormitory effect). Hypothyroidism is a treatable condition but one may wish to talk to the owner about the possibility of heritability of the condition (see Chapter 27). Luteal ovarian cysts may be treated with administration of the luteolytic agent, prostaglandin F2alpha (Lutalyse™; 100 to 250 μg/kg twice daily for 4 days or until serum progesterone declines to less than 1 ng/ml). Side effects include panting, hypersalivation, vomiting, and defecation. These side effects usually subside within 30 to 60 min of treatment. Luteal cysts also can be treated by OHE.

Chromosomal abnormalities reportedly associated with persistent anestrus in bitches include 78,XO; 79,XXX; 79,XXY; and dogs with multiple cell lines. Some bitches with these karyotypes may demonstrate intermittent estrous cycling but all are considered infertile.

In the absence of any defined cause, medical induction of estrus may be attempted. There are no drugs approved for this purpose in bitches in the United States. No drug described can accurately induce estrus on a given date or with complete consistency. Reported regimens are described in Table 51-1.

Table 51-1. Reported estrus induction protocols in dogs.

Drug	Dosing Regimen	Reported Percentage Entering Proestrus	Reported Pregnancy Rate (%)	Comments
Diethylstilbestrol	5 mg once daily for 6 to 9 days, until proestrus induced	100	100	Reported studies had low numbers of animals. May induce split heat (signs of proestrus, no ovulation, return to proestrus with an ovulatory heat about 1 month later).
Bromocriptine	20 μg/kg twice daily per os for 21 days	71 to 100	83	Drug must be procured through a human pharmacy. Readily induces vomiting in dogs.
Cabergoline	5 μg/kg once daily per os until proestrus induced (30 day maximum)	80 to 100	86 to 93	Available as a veterinary drug (Dostinex™). Does not induce vomiting. Anecdotal reports suggest lower percentage of bitches entering proestrus.
Gonadotropin releasing hormone and analogues	—	—	—	Variable success reported with subcutaneous or injectable applications.

Supplemental reading

Cirit U, Bacinoglu S, Cangul IT, et al. 2007. The effects of a low dose of cabergoline on induction of estrus and pregnancy rates in anestrous bitches. *Anim Reprod Sci* 101:134–144.
Kutzler MA. 2007. Estrus induction and synchronization in canids and felids. *Theriogenology* 68:354–374.
Lyle SK. 2007. Disorders of sexual development in the dog and cat. *Theriogenology* 68:338–343.

52

What is the diagnostic approach for infertility of a bitch?

Etiology

The most common cause of apparent infertility in bitches is improper breeding management (see Chapter 29). Other causes include poor semen quality of the male (see Chapter 21), reproductive tract infection including brucellosis (see Chapter 44), systemic disease possibly including hypothyroidism (see Chapter 27), non-receptive behavior due to pain or psychological factors, uterine pathology, anovulatory cycles, impatency of the reproductive tract, and advanced age of the bitch.

Clinical signs

Bitches with infertility may present for failure to stand to be bred, apparent conception failure, or known pregnancy loss. Because there is no early pregnancy test for dogs (see Chapter 5), conception failure and early pregnancy loss cannot be differentiated. Bitches with infection of the reproductive tract or uterine pathology often exhibit no specific clinical signs. Bitches with infertility associated with hypothyroidism rarely show the classical signs of bilaterally symmetrical alopecia and lethargy.

Diagnosis

An accurate history must be taken of any past breeding attempts to determine if the bitch was bred on or near optimal breeding day (see Chapter 29). Reproductive tract infection is best identified by culture of the anterior vagina when the bitch is in estrus (see Chapter 28). Brucellosis serology should be performed (see Chapter 44). Hypothyroidism is best diagnosed by demonstration of decreased free thyroxine and increased thyroid stimulating hormone in serum (see Chapter 27). Behavioral causes are identified by observation of breeding attempts and by digital vaginal examination and vaginoscopy to identify vaginal anatomic anomalies (Table 52-1).

Uterine pathology may or may not be visible ultrasonographically (see Chapter 47, Fig. 47-2). Cystic endometrial hyperplasia (CEH) may be visible as fluffy thickening of the endometrium and presumably interferes with fertility by altering movement of spermatozoa from the cervix to the uterine tubes and interfering with normal implantation and placentation of conceptuses. Uterine pathology is often diagnosed at histopathology. Samples are most commonly retrieved at the time of ovariohysterectomy; uterine biopsy requires laparatomy and hysterotomy and is rarely performed.

Table 52-1. Diagnostic key for infertility of bitches.

1a.	The bitch is cycling.	2
1b.	The bitch is not cycling.	3
2a.	The estrous cycle is normal in length.	4
2b.	The estrous cycle is not normal in length.	5
3a.	The bitch has never been in heat.	6
3b.	The bitch has been in heat earlier in her life.	7
4a.	The male has normal semen quality.	8
4b.	The male does not have normal semen quality.	9
5a.	The estrous cycle is abnormally short.	10
5b.	The estrous cycle is abnormally long.	11
6a.	The bitch is at least 24 months of age.	12
6b.	The bitch is less than 24 months of age.	13
7.	The bitch has secondary anestrus (see above)	
8a.	The bitch is being bred at the right time.	14
8b.	The bitch is not being bred at the right time.	15
9.	See Chapter 56 for a discussion of abnormal semen quality in male dogs.	
10.	Short intervals between estrous cycles are normal in some breeds, most notably the German shepherd dog and Rottweiler. Other causes of abnormally short cycles include uterine disease and split heat.	
11.	Long intervals between estrous cycles may be normal in some breeds, most notably the Basenji, which cycles annually in the fall in the northern hemisphere, and in some lines of dogs. Other causes of abnormally long interestrous intervals include systemic disease and hypothyroidism.	
12.	The bitch has primary anestrus.	
13.	The bitch may be immature. Try housing her with cycling bitches, improving her plane of nutrition, and retiring her from showing or working for a time.	
14.	Possible causes of infertility include reproductive tract infection, hypothyroidism, brucellosis, anovulatory cycles, and genetic incompatibilities with the chosen male.	
15.	Inappropriate breeding management is the most common cause of apparent infertility in bitches (see Chapter 29).	

Impatency of the reproductive tract may be identifiable with hysterography, performed as vaginography when the bitch is in estrus and the cervix open (see Chapter 4). More commonly, abnormalities of the tract are identified during laparatomy. Anovulatory cycles are defined as the cycles in which serum progesterone never rises to the ovulatory range of 4 to 10 ng/ml. In one large survey of bitches, anovulatory cycles were reported to occur in 1% of cycles. Forty-five percent of bitches in that study had a normal, ovulatory cycle following the anovulatory cycle.

Treatment

Ensure the bitch stands to be bred or is inseminated on optimal breeding day, identified using assay of progesterone in serum and adjunct measures (see Chapter 29). Evaluate semen quality of the male if possible or ask for a record of success in recent breedings. Some bitches and males are not compatible genetically; if a bitch has consistently failed to produce pups with one male, recommend use of a male from different lines. Hypothyroidism may be treated but it is most important to educate the owner about potential heritability of the condition in their breed (see Chapter 27). Reproductive tract infection should be treated with appropriate antibiotics, based on culture and sensitivity (see Chapter 28). Some veterinarians recommend surgical insemination for bitches with CEH. At the time of surgery, the uterus is run through the gloved hands of the surgeon and cysts reduced. There are no studies documenting benefit or detriment of this practice on conception rate or general health of the bitch.

Supplemental reading

Grundy SA, Feldman E, Davidson A. 2002. Evaluation of infertility in the bitch. *Clin Tech Small Anim Pract* 17:108–115.

Johnston SD, Root Kustritz MV, Olson PNS. 2001. *Canine and Feline Theriogenology*. Philadelphia: WB Saunders Co., pp. 257–273.

53

How do I treat benign prostatic hypertrophy?

Etiology

Benign prostatic hypertrophy and hyperplasia (BPH) is an age-related phenomenon in male dogs and humans. Testosterone is metabolized to dihydrotestosterone, which is the active compound that stimulates increase in the glandular and connective tissue portions of the prostate. With increasing age, as testosterone secretion slows and relative estrogen secretion increases, prostatomegaly occurs even more quickly, such that the majority of dogs aged 5 to 6 years or more have significant BPH.

Clinical signs

Many dogs with BPH are asymptomatic. The classic clinical sign of BPH is dripping of bloody fluid from the prepuce unassociated with urination. Signs may worsen if the male dog is exposed to a bitch in heat. Other signs are referable to increased size and vascularity of the prostate and include hematuria (as bloody prostatic fluid drains into the urinary bladder), hemospermia, passage of ribbon-shaped stools, and rectal tenesmus. Systemic signs of disease are uncommon in the absence of secondary prostatitis (see Chapter 54).

Diagnosis

The prostate encircles the urethra at the neck of the urinary bladder. It is palpable per rectum until it becomes large enough to pull the bladder forward into the abdomen; eventually it may be palpable per abdomen. The prostate is symmetrical with a distinct median raphe. In dogs with BPH, the prostate remains symmetrical as it increases in size. The dog does not show evidence of pain when pressure is put on the prostate per rectum or when it contracts as the dog ejaculates. Rectal palpation is not a very accurate diagnostic procedure; in one study, BPH was not identified by rectal palpation in 44.8% of affected dogs.

Ultrasonography reveals a uniformly enlarged prostate (Fig. 53-1). Fine-needle aspirate may be performed with ultrasound guidance for collection of samples for cytology or culture (see Chapter 9). Ejaculated prostatic fluid also may be submitted for cytology and culture. In either instance, culture is used to rule out prostatitis (see Chapter 54).

In humans, assay of proteins in serum, for example, prostate specific antigen, can be used to diagnose prostate disease and monitor treatment. These proteins are produced in dogs but are not consistently reflective of health of the prostate gland.

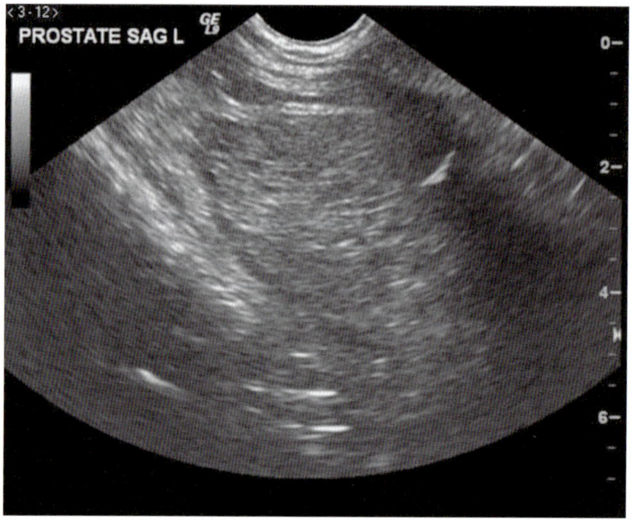

Figure 53-1: Sonogram of canine prostate with benign prostatic hypertrophy.

Treatment

Castration is the best treatment for BPH because this is an androgen-mediated disorder and all significant androgen secretions arise from the testes. No medical therapy is as effective as castration for minimizing clinical signs of BPH. Significant reduction in prostate size occurs within 3 weeks of castration.

For valuable breeding dogs or others that cannot be castrated, several medical therapies are available. Treatment with estrogen compounds has been described historically but is not recommended. Therapy with progestins has also been described (megestrol acetate, 0.5 mg/kg per os once daily for 4 to 8 weeks; medroxyprogesterone acetate, 3 to 4 mg/kg subcutaneously at intervals of 10 weeks or longer). Semen quality is not affected negatively by treatment with progestins.

The preferred medical therapy is with the human drug finasteride (Proscar™—5 mg tablet; Propecia™—1 mg tablet; dose regimen is 5 mg once a day per os for dogs weighing 5 to 50 kg). Finasteride prevents conversion of testosterone to dihydrotestosterone, decreasing prostate size while retaining normal libido and spermatogenesis. No side effects have been reported with use of this drug in dogs. Semen volume decreases as prostate size decreases but conception rates and semen quality do not vary. In human medicine, there is a concern about birth defects; this is not a concern in veterinary medicine since male dogs do not breed bitches once they are pregnant. Saw palmetto and many other human drugs available for treatment of BPH are not useful in dogs.

Supplemental reading

Barsanti JA, Finco DR, Mahaffey MM, et al. 2000. Effects of an extract of Serenoa repens on dogs with hyperplasia of the prostate gland. *Am J Vet Res* 61:880–885.

Freitag T, Jerram RM, Walker AM, et al. 2007. Surgical management of common canine prostatic conditions. *Compendium* 29:656–672.

Jia G, Heverhagen J, Henry H, et al. 2006. Pharmacokinetic parameters as a potential predictor of response to pharmacotherapy in benign prostatic hyperplasia: A preclinical trial using dynamic contrast-enhanced MRI. *Magn Reson Imaging* 24:721–725.

Mukaratirwas S, Chitura T. 2007. Canine subclinical prostatic disease: Histological prevalence and validity of digital rectal examination as a screening test. *J S Afr Vet Assoc* 78:66–68.

Sirinarumitr K, Sirinarumitr T, Johnston SD, et al. 2002. Finasteride induced prostatic involution by apoptosis in dogs with benign prostatic hypertrophy. *Am J Vet Res* 63:495–498.

Smith J. 2008. Canine prostatic disease: A review of anatomy, pathology, diagnosis, and treatment. *Theriogenology* 70:375–383.

54

How do I treat prostatitis?

Etiology

Prostatitis is inflammation of the prostate gland. Usually this occurs secondary to bacterial infection; *Mycoplasma canis* and blastomycosis have also been reported as causative. Causative bacterial organisms are those of the normal urethral flora. The normal canine prostate is built to withstand infection. Infection occurs secondary to prostatic or urinary tract disease. Benign prostatic hypertrophy is the most common underlying problem in intact dogs (see Chapter 53). Prostatic neoplasia, most commonly adenocarcinoma, is the most common underlying problem in castrated dogs.

Clinical signs

Acute prostatitis is associated with clinical signs typical of any acute infection. The dog may be febrile, lethargic, and inappetent. The dog may cry out when ejaculating or refuse to breed. Signs of benign prostatic hypertrophy (BPH) may also be present, including dripping of bloody fluid from the prepuce unassociated with urination, hemospermia, passing of ribbon-shaped stools, and rectal tenesmus. Dogs with prostatic adenocarcinoma usually have metastatic disease by the time prostate disease is evident. Signs may include cachexia, muscle atrophy, respiratory distress, lymphadenopathy, and ataxia.

Chronic prostatitis is associated with fewer systemic signs. Generally, the clinical manifestation is as for the underlying problem. The dog may present for the complaint of infertility or poor semen quality.

Diagnosis

Diagnosis of prostatitis requires demonstration of growth of bacteria from the prostate. Prostatic fluid, collected by ejaculation (see Chapter 6) or prostatic tissue, collected by fine-needle aspirate or biopsy (see Chapter 9) may be submitted for culture and sensitivity. Rectal palpation may be used to determine prostate size and presence of pain upon palpation but is inaccurate for diagnosis; in one study, rectal palpation was used to identify prostatitis in only 23.6% of affected dogs. Ultrasound should be used to guide sample collection. On ultrasound, the infected prostate appears mottled and may have scattered mineralization (Fig. 54-1). Prostatic neoplasia may appear more whorled in appearance. The two conditions cannot be definitively differentiated by ultrasound alone. There are no reported complications from fine-needle aspirate or biopsy of

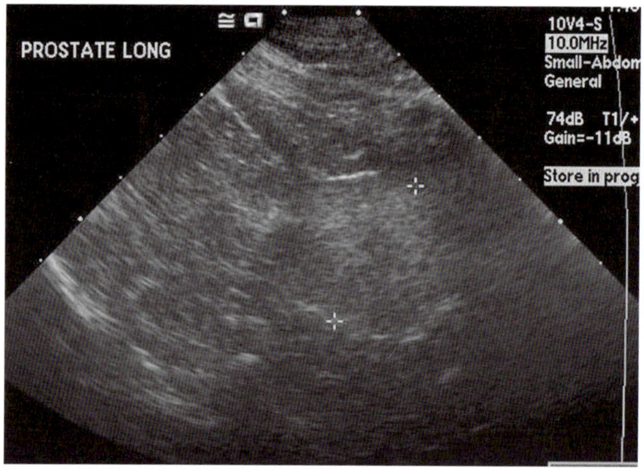

Figure 54-1: Ultrasonogram of canine prostatitis.

the potentially infected prostate. Diagnosis of the underlying disorder is best made by prostatic biopsy, especially if index of suspicion for prostatic neoplasia is high.

Treatment

In acute prostatitis, the prostatic capsule is disrupted and any antibiotic deemed suitable by culture and sensitivity will penetrate the prostatic tissue. In chronic prostatitis, the capsule is intact and only lipophilic antibiotics that are not highly protein bound are likely to penetrate the tissue well. Antibiotics most efficacious for treatment of chronic prostatitis include fluoroquinolones, trimethoprim-sulfa, and chloramphenicol. Long-term use of trimethoprim-sulfa drugs may be associated with anemia and keratoconjunctivitis sicca. Treatment with the antibiotic chosen should last for 4 to 6 weeks. Reculture prostatic tissue or fluid 1 week and again 1 month after therapy and retreat if necessary. Concurrent treatment for BPH with finasteride or castration may hasten resolution. If prostatic adenocarcinoma is present, the animal should be treated palliatively.

Supplemental reading

Freitag T, Jerram RM, Walker AM, et al. 2007. Surgical management of common canine prostatic conditions. *Compendium* 29:656–672.

L'Abee-Lund TM, Heiene R, Friis NF, et al. 2003. Mycoplasma canis and urogenital disease in dogs in Norway. *Vet Rec* 153:231–235.

Mukaratirwa S, Chitura T. 2007. Canine subclinical prostatic disease: Histological prevalence and validity of digital rectal examination as a screening test. *J S Afr Vet Assoc* 78:66–68.

Smith J. 2008. Canine prostatic disease: A review of anatomy, pathology, diagnosis, and treatment. *Theriogenology* 70:375–383.

Wagenlehner FME, Weidner W, Sorgel F, et al. 2005. The role of antibiotics in chronic bacterial prostatitis. *Int J Antimicrob Agents* 26:1–7.

55

What is the best treatment for recurrent paraphimosis (extrusion of the penis from the prepuce) in neutered male dogs?

Etiology

Paraphimosis occurs most commonly in older, castrated toy breeds, although it may occur in any breed. Underlying causes are medical (balanoposthitis [infection of the penile and preputial mucosa], foreign object, neurologic disease) or behavioral. Erection of the penis is not hormonally mediated but is instead neurologic and so can occur in dogs regardless of intact status. Any irritation of the penile mucosa or of the nerves mediating erection and protrusion of the penis may cause drying of the penis and inability of the dog to retract the penis completely into the prepuce.

Balanoposthitis is a secondary infection with the normal preputial flora. Underlying causes include foreign objects, prostate disease, and atopic dermatitis.

Behavioral causes of paraphimosis include unintended positive reinforcement and sequencing. Many owners very vigorously scold the dog for extruding or licking at the penis; dogs may see this increased attention as a positive reinforcement. Similarly, dogs may learn a sequence of events (go out to urinate, lick at the penis, get a treat) and exhibit the behavior for a reward.

Clinical signs

Paraphimosis is extrusion of the non-erect penis from the prepuce. This should be differentiated from priapism, which is persistent erection of the penis and which is associated with systemic effects and eventual nonviability of penile tissue. In dogs with paraphimosis, the exposed portion of the penis dries and reddens, especially if chronically exposed. The dog does not usually feel pain when the area is examined. If balanoposthitis is present, creamy, yellowish discharge overlies the penis when extruded and the penile and preputial mucosa is erythematous. Neurologic disease may be associated with other neurologic signs.

Diagnosis

Extrusion of the penis and inspection of the prepuce are diagnostic for balanoposthitis and presence of foreign objects. A sample of preputial discharge should be submitted for culture and sensitivity. Rectal examination of the prostate should be performed. In castrated dogs, the prostate should barely be palpable. Prostatomegaly in a castrated dog is highly suspicious of prostatic neoplasia, and carries a poor prognosis. The dog should be evaluated for atopic dermatitis. A complete neurological examination should be performed. Owners should be questioned about the frequency with which the problem occurs and about any possible sequencing leading to or reinforcing this behavior.

Treatment

Balanoposthitis is treated with appropriate antibiotic therapy, based on culture and sensitivity, and treatment of the underlying cause, if identified. Behavioral causes should be addressed by teaching the owner not to overreact when disciplining the dog for the behavior and to try to break up sequences leading to the behavior. It may be beneficial to leave a leash on the dog when in the house so when he starts licking, the owner can just gently drag his head away from that area and distract him.

If no cause can be identified, some dogs respond to therapy with progestins, which have anti-inflammatory and antianxiety properties. Examples include megestrol acetate (0.5 mg/kg per os once daily for a maximum of 30 days) and medroxyprogesterone acetate (2.5 mg/kg subcutaneously every 5 months with a maximum of two treatments). Side effects of progestins in male dogs are polyphagia, diabetes mellitus, and possibly mammary neoplasia.

Some dogs self-traumatize to a large extent. These dogs may be best treated with antipsychotic drugs or tranquilizers. Surgical replacement and fixation of the penis has been described, with suturing of the dorsal penile epithelium to the apposing preputial surface.

Supplemental reading

Root Kustritz MV. 2009. Intermittent erection of the penis in castrated male dogs. In *Kirk's Current Veterinary Therapy XIV* (Bonagura JD, Twedt DC, and Kirk RW, eds.), p. 1053. St. Louis, MO: Elsevier.

Somerville ME, Anderson SM. 2001. Phallopexy for treatment of paraphimosis in the dog. *J Am Anim Hosp Assoc* 37:397–400.

56

What is the diagnostic approach for infertility of a male dog?

Etiology

Infertility may be due to problems with the bitch (see Chapter 52) or the male. Causes of infertility described in male dogs include inability to breed and poor semen quality. Causes of inability to breed include poor libido and pain, as may be seen in dogs with prostate disease. Causes of poor semen quality include pre-testicular causes, such as systemic disease including hypothyroidism (see Chapter 27) and immotile cilia syndrome; testicular causes including testicular atrophy, testicular neoplasia, and brucellosis (see Chapter 44); and post-testicular causes including spermatocele and retrograde ejaculation.

Clinical signs

Poor libido is manifested by a male dog expressing no interest in mounting an estrous bitch. Pain may be evidenced in the same way or may be seen as mounting followed by quick dismount or even crying out as the dog ejaculates. Prostate disease may be evidenced by dripping of bloody fluid from the prepuce unassociated with urination, hematuria, hemospermia (blood in the ejaculate), presence of ribbon-shaped feces, or pain when ejaculating (see Chapters 53 and 54). Immotile cilia syndrome, also termed ciliary dyskinesia, includes lack of motility in the respiratory tract and so is usually evidenced by chronic respiratory disease. Hypothyroid dogs may or may not exhibit classic signs of bilaterally symmetrical alopecia and lethargy. Testicular atrophy is self-evident and may also be seen as a component of testicular neoplasia, especially if the contralateral testis is enlarged. Brucellosis is described in Chapter 44.

Diagnosis

Poor libido may be due to pain or may be behavioral in origin. Sources of pain include the caudal spine, rear limbs, and prostate. Arthritis or other causes of bone pain are visible radiographically. Diagnosis of prostate disease is described in Chapters 53 and 54. Definitive diagnosis of immotile cilia syndrome requires electron microscopy of immotile spermatozoa or cells retrieved from the respiratory tract. Behavioral causes include frequent disciplining when having shown mounting behavior in the past, or recollection of pain when collected or bred in the past (Table 56-1).

Poor semen quality may be defined as aspermia (lack of an ejaculate), azoospermia (ejaculation of seminal fluid containing no spermatozoa), oligozoospermia (ejaculation of an abnormally low number of spermatozoa), teratozoospermia (ejaculation of an abnormally low percentage

Table 56-1. Diagnostic key for infertility in male dogs.

1a.	The dog is capable of breeding bitches naturally (normal mount and erection).	2
1b.	The dog is not capable of breeding bitches naturally.	3
2a.	Semen quality is normal.	4
2b.	Semen quality is not normal.	5
3.	Possible causes include psychological constraints, pain in the rear limbs or spine, prostate disease, and abnormalities of the penis.	
4a.	The bitch is at the correct stage of her estrous cycle for breeding.	6
4b.	The bitch is not at the correct stage of her estrous cycle for breeding.	7
5a.	Possible causes for abnormal semen quality are brucellosis, hypothyroidism, testicular atrophy, testicular neoplasia, infection of the reproductive tract, prostate disease, retrograde ejaculation, and immotile cilia syndrome.	
6b.	Possible causes include prostate disease, brucellosis, testicular neoplasia, and genetic incompatibility with the chosen female.	
7.	Improper breeding management is the most common cause of apparent infertility in bitches (see Chapters 29 and 52).	

of normally shaped spermatozoa), or asthenozoospermia (ejaculation of abnormally low percentage of progressively motile spermatozoa). These abnormalities often occur together.

Aspermia is often due to pain. Sources of pain include the caudal spine, rear limbs, and prostate. Arthritis or other causes of bone pain are visible radiographically. Diagnosis of prostate disease is described in Chapters 53 and 54.

Azoospermia may be due to pain or apprehension, such that the dog does not provide a complete ejaculate, may be indicative of lack of spermatogenesis, or may be due to obstruction of the ductus deferens preventing movement of spermatozoa from the epididymis through the urethra. Alkaline phosphatase (ALP) is secreted into seminal fluid from the epididymes and testes. Measurement of ALP in seminal fluid can be used to assess if a complete ejaculate was provided. If concentration of ALP in azoospermic seminal fluid is greater than 5000 IU/l, a complete ejaculate was obtained and the dog is not making spermatozoa. If concentration is lower than 5000 IU/l, a complete ejaculate was not obtained. Semen collection should be attempted again with all distractions removed (see Chapter 6) and diagnostics performed as above to identify sources of pain. Blockages along the spermatic cord, such as spermatocele, may be identified by a skilled ultrasonographer.

Oligozoospermia may be due to incomplete ejaculation, retrograde ejaculation, or reproductive tract disease. Some reports suggest that dogs with hypothyroidism may have abnormally low sperm numbers; hypothyroidism is best diagnosed by demonstration of abnormally low free thyroxine and abnormally high thyroid stimulating hormone (see Chapter 27). Retrograde ejaculation is diagnosed by collection of a urine sample by cystocentesis immediately after ejaculation and demonstration of a greater number of spermatozoa in the urine than in the seminal fluid. Testicular changes that may be evident on palpation include increase in testicular size, most often due to neoplasia, and decreased testicular size, most often due to atrophy. The three common types of testicular neoplasia are seminoma, interstitial (Leydig) cell tumor, and Sertoli cell tumor, which may secrete estrogen and be associated with bilaterally symmetrical alopecia, attraction of male dogs, gynecomastia, and atrophy of the contralateral testis. Metastasis is uncommon with all three tumor types. Some testicular tumors are not palpable and are best identified by ultrasound. Testicular atrophy may occur secondary to testicular neoplasia in the atrophied or contralateral testis, elevated body temperature due to fever or heat stroke, or direct increase in scrotal temperature, as may occur with trauma or frostbite. Oligozoospermia may be a manifestation of prostate disease (see Chapters 53 and 54) or brucellosis (see Chapter 44).

Teratozoospermia and asthenozoospermia often occur together and are usually due to the same causes as oligozoospermia. Asthenozoospermia in the presence of a large number of normally shaped spermatozoa may be indicative of contaminated equipment; another semen sample should be collected using clean equipment and meticulous care should be taken not to expose the semen to soap or other toxic substances.

Treatment

Poor libido due to pain should be treated by management of the underlying cause, if possible. Undefined causes of poor libido may be treated by administration of gonadotropin releasing hormone (1 to 2 µg/kg intramuscularly 60 min before semen collection or breeding). The author has had some success with apprehensive dogs by spraying her lab coat with dog-appeasing pheromone several minutes before collection.

Treatment for prostate disease is discussed in Chapters 53 and 54, for brucellosis in Chapter 44, and for hypothyroidism in Chapter 27. Castration cures most cases of testicular neoplasia.

Retrograde ejaculation may be treated with sympathomimetic agents, which promote closure of the bladder neck during ejaculation. Examples include phenylpropanolamine (3 mg/kg per os twice daily) and pseudoephedrine hydrochloride (3 to 5 mg/kg per os three times daily or 3 and 1 h before semen collection or attempted breeding). If no specific cause of oligozoospermia is identified, number of spermatozoa in the ejaculate may be increased by administration of prostaglandin F2alpha (Lutalyse™, 0.1 mg/kg subcutaneously 15 min prior to collection). Side effects are mild and include salivation and panting lasting 10 to 20 min.

Supplemental reading

Hess M. 2006. Documented and anecdotal effects of certain pharmaceutical agents used to enhance semen quality in the dog. *Theriogenology* 66:613–617.

Memon MA. 2007. Common causes of male dog infertility. *Theriogenology* 68:322–328.

Romagnoli S, Majolino G. 2009. Aspermia/oligozoospermia caused by retrograde ejaculation in the dog. In *Kirk's Current Veterinary Therapy XIV* (Bonagura JD, Twedt DC, and Kirk RW, eds.), pp. 1049–1053. St. Louis, MO: Elsevier.

Section V

Feline techniques

57

What is the technique for collection of a vaginal cytology specimen?

FELINE TECHNIQUES

Anatomy

The feline vagina is short and horizontal, with a significant narrowing less than 1 cm within the vulvar lips.

Pre-procedure considerations

The vagina is not sterile. A non-sterile cotton-tipped applicator, moistened with saline or water, is used. Smaller diameter swabs may be used, but the operator must be careful of their flexible shaft.

Procedure

Estrous queens readily tolerate the vaginal swab procedure. Anestrous queens do not tolerate the procedure and must be restrained. Insert the moistened swab between the vulvar lips to the depth of the cotton. Roll the swab within the vagina, remove it, and roll it several times on a clean glass slide. Allow the slide to air-dry and stain for interpretation (see Chapter 58).

Post-care and complications

No complications are reported.

Supplemental reading

Watson PF, Glover TE. 1993. Vaginal anatomy of the domestic cat (Felis catus) in relation to copulation and artificial insemination. *J Reprod Fert* 47(Suppl):355–359.

Zambelli D, Cunto M. 2005. Vaginal and cervical modifications during the estrus cycle in the domestic cat. *Theriogenology* 64:679–684.

58

How do I interpret vaginal cytology?

Anatomy

Not applicable.

Pre-procedure considerations

Not applicable.

Procedure

Breeding management

The healthy epithelial cells that line the vagina at all times are termed non-cornified and include the parabasal and intermediate cells (Fig. 58-1). Under the influence of estrogen, these cells are

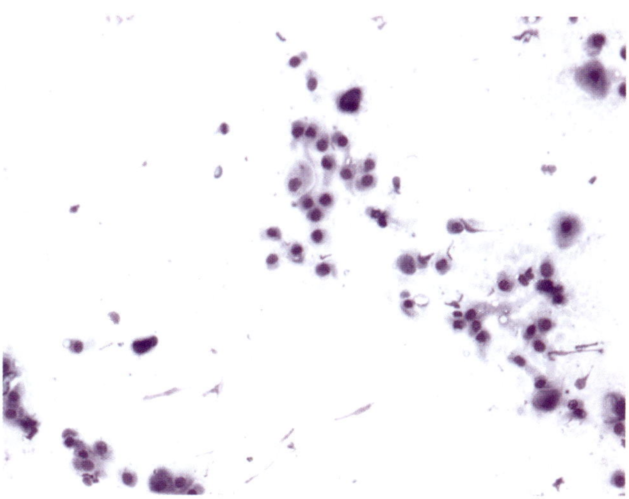

Figure 58-1: Non-cornified vaginal epithelial cells from a queen.

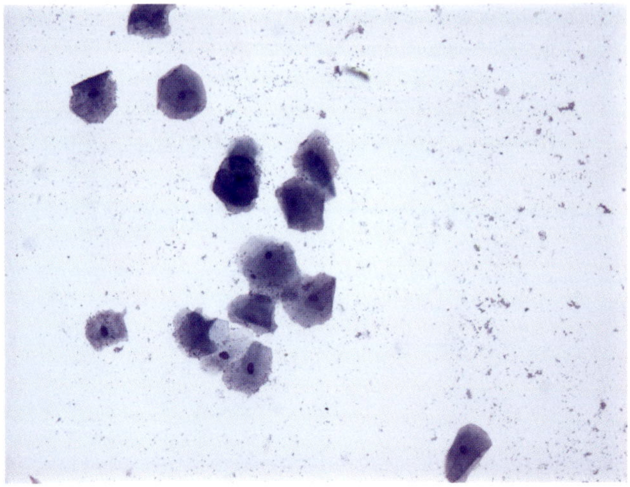

Figure 58-2: Cornified vaginal epithelial cells from a queen.

stimulated to divide. As the cells divide and the vaginal lining thickens, those cells nearest the lumen become non-viable and lose the characteristic appearance of a healthy cell monolayer. The misshapen, clumped cells are termed cornified (Fig. 58-2). Cornified vaginal cytology in queens varies from that in bitches in that the cornified cells are less likely to clump together and rarely appear anuclear.

Diagnosis of disease

Spayed female cats and queens not in proestrus or estrus will have non-cornified epithelial cells. Presence of cornified cells at any other time suggests estrogen influence. Possible disorders associated with abnormally cornified cytology include ovarian remnant syndrome, ovarian cystic disease, ovarian tumor, or ingestion or absorption of estrogen, usually from a human pharmaceutical source. Pathologic discharge usually is mucopurulent and is associated with uterine disease (see Chapter 80) or pregnancy loss (see Chapter 79).

Post-care and complications

Not applicable.

Supplemental reading

Johnston SD, Root Kustritz MV, Olson PNS. 2001. *Canine and Feline Theriogenology*. Philadelphia: WB Saunders Co., pp. 396–405.

59

What techniques are available for pregnancy diagnosis and when are they best used?

Anatomy

The uterine body lies in the caudal abdomen and is the only discrete tissue mass between the colon and the urinary bladder. The uterine horns are tortuous and lie cranially among the small intestines.

Pre-procedure considerations

Pregnancy diagnosis in cats is easy. Because of the small size of queens, there may be a greater risk of harm to the embryo or fetus with manipulation of the amniotic sac or conceptus. The uterine horns are tortuous and lie among the small intestines. The stage at which pregnancy can be diagnosed varies with the technique used. There are few reasons to see potentially pregnant cats outside this window, and, in fact, this may expose them to disease or otherwise potentially harm the pregnancy.

Procedure

Diagnostic tests valuable for pregnancy diagnosis in queens and the time they are best used are listed in Table 59-1. There is no early pregnancy test for cats, and human early pregnancy tests do not work in cats because they assay a hormone produced only by humans, human chorionic gonadotropin.

Post-care and complications

There are no reported detriments to the queen or kitten with any of these diagnostic techniques if performed at the proper time with technical skill and use of well-maintained equipment.

Supplemental reading

Harris LA, Steinetz BG, Bond JB, et al. 2008. Refinement of a commercial bench-top relaxin assay for pregnancy diagnosis using urine from domestic and nondomestic felids. *J Zoo Wildl Med* 39:10–179.

Johnston SD, Root Kustritz MV, Olson PNS. 2001. *Canine and Feline Theriogenology*. Philadelphia: WB Saunders Co., pp. 414–430.

Table 59-1. Pregnancy diagnostic techniques in the queen.

Technique	Time When Best Used	Comments	Litter Size?	Viability?
Abdominal palpation	21–35 days of pregnancy	Prior to 21 days, the individual amniotic vesicles are difficult to feel, and after 35 days, they enlarge and become confluent, again making them difficult to feel as individual entities. Care should be taken not to squeeze the amniotic vesicles but rather to let them "blip" through the fingers. Palpation is difficult in obese and tense queens.	Palpation is a poor indicator of litter size.	Palpation cannot be used to assess viability of kittens. At term, kitten movement may be visible or palpable, but lack of movement is not invariably associated with kitten death.
Transabdominal ultrasonography	Beyond 16 days of pregnancy	At 16 days, the amniotic vesicles are visible as black balls with a comma-shaped tissue mass within them. Beyond 16 to 25 days, beating hearts can be seen.	Ultrasonography is a poor indicator of litter size unless the litter is very small and the kittens are widely spaced in the abdomen.	Ultrasonography is the best indicator of viability. Beyond assessment of movement, heart rate can be assessed; heart rate of less than 150 to 170 bpm is indicative of fetal stress.
Abdominal radiographs	Beyond 38 to 40 days of pregnancy	Mineralization of fetuses must be present. Radiographs are most useful very late in pregnancy, when not only can pregnancy be diagnosed, but litter size, size of individual kittens, and some notion of viability also may be elucidated.	Radiography is the best indicator of litter size. Miscounts are most common in very large litters.	If kittens have been dead for at least 1 day, signs of fetal death may be visible on radiographs, including gas within and around the kittens, and collapse of the skull or axial skeleton.
Relaxin assay	Beyond 28 days of pregnancy	The only in-house assay that is commercially available is marketed for dogs. This test can be used to assay relaxin in cat serum or in cat urine that is diluted 1 : 1 with nonpregnant cat serum.	—	—

60

What are the techniques for semen collection from male cats?

Anatomy

Intact male cats have a penis covered with cornified spines, two testes, which should descend into the scrotum shortly after birth, a prostate, and bulbourethral glands.

Pre-procedure considerations

Manual ejaculation requires training and may not be possible for all cats, especially those who will be seen only occasionally in practice. Electroejaculation requires general anesthesia; pre-anesthetic physical examination and bloodwork should be performed.

Procedure

Manual ejaculation is as in the dog. A tiny artificial vagina may be created using a 2-ml rubber pipette bulb and a 3 × 44 mm test tube. The cat is trained by stimulation in the presence of an estrous queen. The collecting instrument is placed over the penis as the cat ejaculates. This technique requires training of the operator and the cat, and is not possible in all cats. Normal breeding behaviors of male cats are aggressive and operators should exercise great caution.

Electroejaculation requires the use of a probe specifically sized for domestic cats, and a rheostat allowing the application of stimuli varying from 0 to 5 V (Fig. 60-1). Stimuli are applied in an increasing sequence with rest periods between, and a collecting device placed over or beneath the penis of the recumbent cat to catch the ejaculate.

Miscellaneous techniques that permit evaluation of whether spermatozoa are being ejaculated or not, but do not provide a volume of semen for evaluation, are cystocentesis of the tom after natural service and collection of a vaginal swab from the queen after natural service. Lack of spermatozoa with either technique is not conclusive evidence that ejaculation did not occur or that the tom ejaculated fluid containing no spermatozoa; presence of spermatozoa with either technique confirms that the male is producing and ejaculating some spermatozoa.

Post-care and complications

Recovery from anesthesia is as for any procedure. The most common side effects reported for electroejaculation impact semen quality only and do not affect the health of the cat. Semen quality may be decreased when semen is collected by electroejaculation as overstimulation of the accessory sex glands increases ejaculation of sperm-free fractions of the ejaculate, and as

Figure 60-1: Electroejaculation equipment for domestic cats (photo courtesy of Marie Kustritz).

retrograde movement of spermatozoa into the urinary bladder may occur. The latter effect may be more significant if xylazine is used as an anesthetic agent. Concentration of spermatozoa in the ejaculate has been demonstrated to be increased when medetomidine was compared with ketamine for induction of anesthesia.

Supplemental reading

Johnston SD, Root Kustritz MV, Olson PNS. 2001. *Canine and Feline Theriogenology*. Philadelphia: WB Saunders Co., pp. 508–522.

Zambelli D, Cunto M, Prati F, Merlo B. 2007. Effects of ketamine or medetomidine administration on quality of electroejaculated sperm and on sperm flow in the domestic cat. *Theriogenology* 68:796–803.

61

What is the technique for semen evaluation in cats?

Anatomy

Not applicable.

Pre-procedure considerations

Cold shock is not as significant a problem in cats as it is in other species. Equipment and samples can be maintained at room temperature for evaluation.

Procedure

Color

Color is assessed visually. Normal semen is moderately milky white. Abnormal colors that may be seen include clear (no spermatozoa in the ejaculate) and yellow (urine contamination).

Volume

Volume varies in cats from 0.01 to 0.7 ml. Volume is not correlated with quality. Record volume collected before any samples are removed for evaluation; this value will be needed later to calculate the total number of spermatozoa in the ejaculate.

Motility

Motility should be assessed soon after semen collection. Place one drop of semen on a glass slide. You may or may not use a cover slip. Subjectively assess the percentage of spermatozoa that are moving forward. Normal is 70% or greater. Some people also assess the speed of movement of the spermatozoa; there are no reported correlations between speed of motility and fertility in that cat or with use of that particular sample.

Concentration/total number

Make a 1:100 dilution of semen by diluting 1 part semen (0.1 ml) with 9 parts formal-buffered saline (0.9 ml) to make a 1:10 dilution and than mixing 1 part of that initial dilution with 9 parts formal-buffered saline. This also can be done using the white blood cell Unopette system. Use the enclosed piercing device to make a hole in the top of the diluent container. Remove the piercing device to reveal the pipette. Fill the pipette by capillary action. Squeeze the diluent container and insert the pipette, letting go of everything at once so the semen is sucked from the pipette into the diluent. Remove the pipette and turn it around, reseating the hub to form a dispenser.

With either technique, semen is counted using a hemacytometer. Place the glass cover slip over the central area of the hemacytometer. Dispense diluted semen such that capillary action carries it across the central area. Fill each side of the hemacytometer separately. Allow the hemacytometer to sit for about 5 min after filling, to allow spermatozoa to settle.

The hemacytometer grid consists of nine large squares. Using the 10× objective, one of these large squares will fill the microscope field. Count all the spermatozoa visible in one of the nine large squares. This yields the concentration in millions per milliliter.

Total number of spermatozoa in the sample does not vary with technique and is the valued number in semen evaluation. Volume (milliliter per ejaculate) multiplied by concentration (spermatozoa per milliliter) yields total number (spermatozoa per ejaculate). Normal ranges from 3.5 to 61 million spermatozoa per ejaculate.

Morphology

Morphology, or shape, of spermatozoa is not specifically associated with fertility in cats as in other species. However, a higher percentage of morphologically normal spermatozoa is desirable as poor morphology often is associated with poor motility and, presumably, with decreased ability of those spermatozoa to reach the uterine tube and fertilize ova.

Morphology can be assessed in wet mounts using phase-contrast microscopy, if it is available. Head defects may be more difficult to see with this technique, and it is not available to most practitioners. It may be more difficult to see other abnormalities, most notably cytoplasmic droplets, on stained samples.

Place one drop of semen at one end of a glass slide and smear it out as for a blood smear. Let it air-dry and stain with a triple stain, leaving it in each of the three solutions (fixative, safranin, and crystal violet) for 5 min. Rinse and allow to air-dry before evaluating under oil immersion.

Another technique is to place one drop of semen at one end of a glass slide and place a drop of an eosin-nigrosin stain next to it. Using a pusher slide, rock the 2 drops together to mix them and draw it out into a thick smear. Allow to air-dry and evaluate under oil immersion.

Morphology can be broken down into normal spermatozoa, those with primary defects, and those with secondary defects. Primary defects occur during spermatogenesis and include anything doubled, any abnormality of the shape of the head, proximal cytoplasmic droplets, and bending of the midpiece. Secondary defects occur during sample preparation or are an indication of infection and include detached heads, distal cytoplasmic droplets, and bent tails. The average percentage of morphologically normal spermatozoa in one report was 44%; other sources cite values nearer 71%. The significance of specific primary or secondary defects is not known in cats, although a preponderance of primary defects may carry a worse prognosis than a preponderance of secondary defects, simply because the underlying cause is less likely to be something easily addressed. Percentage of secondary defects has been demonstrated to change from one collection to the next in cats, suggesting that multiple evaluations be performed before a cat is declared subfertile or infertile.

Use of automated semen evaluation systems

Many facilities that do much semen evaluation, especially those that freeze semen, use an automated system for semen evaluation. These are commonly called computer-assisted sperm analysis (CASA) systems. CASA systems must be specifically calibrated for cat semen, which is more dilute than that of other species.

Post-care and complications

Not applicable.

Supplemental reading

Axner E, Linde-Forsberg C. 2007. Sperm morphology in the domestic cat and its relation with fertility: A retrospective study. *Reprod Dom Anim* 42:282–291.

Johnston SD, Root Kustritz MV, Olson PNS. 2001. *Canine and Feline Theriogenology*. Philadelphia: WB Saunders Co., pp. 508–520.

62

What is the technique for vaginal insemination?

Anatomy

The feline vagina is small and horizontal, with a significant narrowing less than 1 cm within the vulvar lips.

Pre-procedure considerations

Ovulation must be induced before insemination is performed. Administer 100 to 250 IU human chorionic gonadotropin intramuscularly to the estrous queen days 2 to 4 after onset of behavioral estrus. Insemination usually is performed 12 to 24 h later.

Vaginal insemination is used with fresh semen. Frozen–thawed semen must be inseminated directly into the uterus (see Chapter 63).

Procedure

Total insemination volume should be about 0.1 ml. Semen collected may need to be suspended in an extender to achieve this volume. Place the queen under general anesthesia. Insert either a 1.5-mm diameter silver abscess cannula or bulb-tipped 20-gauge needle into the vagina. Connect a 1-ml syringe containing the semen sample and dispense it into the vagina.

Post-care and complications

Post-care is as for any procedure requiring general anesthesia.

Supplemental reading

Chatdarong K, Axner E, Manee-In S, et al. 2007. Pregnancy in the domestic cat after vaginal or transcervical insemination with fresh or frozen semen. *Theriogenology* 68:1326–1333.

Tanaka A, Takagi Y, Nakagawa K, et al. 2000. Artificial intravaginal insemination using fresh semen in cats. *J Vet Med Sci* 62:1163–1167.

Villaverde AISB, Melo CM, Martin I, et al. 2008. Comparison of efficiency between two artificial insemination methods using frozen-thawed semen in domestic cat (Felis catus). Artifical insemination in domestic cats. *Anim Reprod Sci* doi:10.1016/j.anireprosci.2008.10.008.

63

What is the technique for surgical insemination?

Anatomy

The uterus of the queen has a small body and long, torturous horns. Semen can be deposited anywhere within the lumen of either horn.

Pre-procedure considerations

Surgical insemination is used to introduce semen directly into the uterus, bypassing the vagina and cervix. It most commonly is used with frozen–thawed semen. Ovulation must be induced before insemination is performed. Administer 100 to 250 IU human chorionic gonadotropin intramuscularly to the estrous queen days 2 to 4 after onset of behavioral estrus. Insemination usually is performed 12 to 24 h later.

Procedure

Place the animal under general anesthesia and shave and prepare the abdomen for sterile surgery. Make a ventral midline incision and exteriorize the uterus. Pass a 25-gauge needle or catheter into the lumen of the uterine body.

If frozen semen is being used, semen should be thawed using the instructions provided by the person who froze that semen. Semen should be thawed as close to the moment of insemination as possible to minimize the time the thawed semen is held at room temperature.

Draw the semen up into a sterile syringe. Introduce the semen through the catheter; it should flow freely into the uterine lumen. Even small volumes distend the uterine body and both horns. Muscle fasciculations of the uterus may be evident after the semen is introduced. Withdraw the needle or catheter, and blot the hole with gauze until hemostasis is achieved. Replace the uterus in the abdomen and close in a routine manner.

Post-care and complications

Incision care is as for any abdominal incision.

Supplemental reading

Johnston SD, Root Kustritz MV, Olson PNS. 2001. *Canine and Feline Theriogenology*. Philadelphia: WB Saunders Co., pp. 406–413.

64

What is the technique for anesthesia for Cesarean section?

Anatomy

Anatomic changes of pregnancy are reflected in physiologic changes, as described below.

Pre-procedure considerations

Physiologic changes of pregnancy that impact anesthesia include an increase in plasma volume and decrease in hematocrit, increased oxygen consumption and carbon dioxide production, increased cardiac output, decreased blood pressure due to vasodilation and hypotension due to mechanical compression from the gravid uterus, and slowed gastrointestinal (GI) motility. These changes necessitate addressing hydration to maintain normal blood pressure during surgery, and use of a well-fitted, cuffed endotracheal tube to counter aspiration of GI contents.

Once a decision has been made to perform Cesarean section, time is of the essence. An intravenous catheter should be placed and fluid therapy instituted. The surgical site should be clipped and prepped as much as possible before any anesthetic agents are introduced. Equipment for neonatal resuscitation should be assembled, including oxygen and small masks or endotracheal tubes, hemostats to clamp off the umbilical cords, suction apparatuses to remove fluid from the oral and nasal cavities, warm towels, and appropriate medications (see below).

Procedure

Considerations of anesthesia

Maternal stability under anesthesia should be monitored by assessment of heart rate and rhythm, blood pressure, and oxygenation status (pulse oximetry), if possible. Maintaining maternal blood pressure is an important component in maintaining placental blood flow and supporting the kittens until they are delivered. Adequately oxygenate the queen at all times.

Choosing appropriate anesthetic agents

Remember that most anesthetic agents exert their effect because they are soluble in fat and move readily into tissue. That means that most will very readily cross the placenta and that anything given to the dam is also being given to the kittens. Choose drugs that cause minimal cardiovascular depression, or that are easily reversed. Short-acting drugs will have less of an effect on the fetuses as they are more rapidly cleared; longer acting drugs can be added for the sake of the dam after the kittens are delivered.

Pregnant animals have an increased pain threshold due to release of endogenous endorphins. When considering anesthetic agents, generally this translates to a reduction in inhalant anesthesia requirements and reduced need for opioids to achieved desired analgesia. Sample protocols are outlined in Table 64-1.

Table 64-1. Sample anesthetic protocols for feline Cesarean section.

Protocol 1—If the queen is healthy

- Pretreat with glycopyrrolate
- Induce anesthesia with diazepam and ketamine
- Bolus ketamine intravenously (IV) as needed for intubation and maintenance on isoflurane or
- Use 0.5–1% lidocaine without epinephrine along incision line—Do not exceed total dose of 10 mg/kg of lidocaine

Protocol 2—If the queen is depressed (this protocol may cause excitation in healthy queens)

- Butorphanol or oxymorphone
- Add ketamine or thiopental for inducation; intubate and maintain on isoflurane
- Sedation in kittens can be reversed with naloxone (for opioids) or flumazenil (for diazepam)

Protocol 3

- Propofol for induction; intubate and maintain on isoflurane or sevoflurane

Protocol 4

- Diazepam (intramuscularly) and etomidate (IV); intubate and maintain on isoflurane or sevoflurane

Table 64-2. Kitten resuscitation.

1. Keep the kitten warm. Dry by vigorously rubbing the kitten with a warm, dry towel. Clear the mouth and nose of fluid by suction.
2. If opioids were given to the dam, reverse with naloxone (one drop from a 25-gauze needle, sublingually). If benzodiazepines were given to the dam, reverse with flumazenil.
3. If the neonate is breathing at least 10 bpm and moving or crying, finish drying the kitten, place it in a warm area (warm air circulating pad or radiant heat), and clamp or tie off the umbilicus. If the kitten is not breathing at least 10 bpm, and is also not moving or crying, provide oxygen for 30–40 s and continue tactile stimulation by rubbing.
4. If the kitten does not begin to breathe spontaneously, attempt to stimulate respiration with placement and rotation of a 25-gauge needle into the nasal philtrum (GV 26 or Jen Chung acupuncture site). If that does not work, consider administration of doxapram sublingually.
5. If there is no discernible heart rate, begin gentle chest compressions (1 to 2 per second).
6. If oxygenation and chest compressions do not elicit spontaneous breathing and heart beat, consider trying placement of an intraosseous catheter (see Chapter 90) and administration of intravenous 10% dextrose solution (2–4 ml/kg as a slow bolus) or sodium bicarbonate (1 ml/kg of the 1 IU/ml concentration).
7. If there is no response by 30 min, the kitten is declared dead.

Post-care and complications

The queen is recovered from anesthesia and extubated in a routine fashion.

Steps that can be taken to enhance success in kitten resuscitation include (1) appropriate selection of anesthetic agents, as described above, and (2) implementation of a sequence of steps to be taken by all personnel resuscitating kittens (Table 64-2).

Supplemental reading

Pascoe PJ, Moon PF. 2001. Periparturient and neonatal anesthesia. *Vet Clin NA* 31:315–341.

Raffe MR, Carpenter RE. 2007. Anesthetic management of Cesarean section patients. In *Lumb and Jones' Veterinary Anesthesia and Analgesia* (Tranquilli WJ, Thurmon JC, Grimm KA, eds.), pp. 955–967. Ames, IA: Blackwell Publishing.

Skarda RT. 1999. Anesthesia case of the month: Dystocia in a queen. *J Amer Vet Med Assoc* 214:37–39.

Traas A. 2008. Resuscitation of canine and feline neonates. *Theriogenology* 70:343–348.

Tranquilli WJ. 1992. Anesthesia for Cesarean section in the cat. *Vet Clin NA* 22:484–486.

65

What is the technique for Cesarean section?

Anatomy

The ovaries lie caudal to the kidneys and each is encased in an ovarian bursa. The primary blood supply is the ovarian artery. The body of the uterus lies within the pelvis; the uterine horns lie in the abdomen. The uterus and ovaries are suspended and attached to the body wall by the broad ligament. The primary uterine blood supply is the bilateral uterine arteries, which arise from the vaginal arteries caudally and lie along the uterine body with small branches supplying the uterine horns over their length.

Pre-procedure considerations

Cesarean section (C-section) is indicated if obstructive dystocia is present, if the queen is exhibiting primary or secondary uterine inertia, if medical therapy for dystocia has been unproductive, or if fetal heart rate is less than 170 bpm (see Chapter 78). Concurrent ovariohysterectomy (OHE) should be offered to the owner. The primary advantage is that the queen will not have to undergo another anesthetic episode for OHE in the future. The disadvantages are an increased risk of hemorrhage and hypovolemia. Milk production will not be altered by concurrent OHE.

The queen should be prepared for surgery to the greatest extent possible before any anesthetic medications are administered. Details of anesthesia are presented in Chapter 64.

Procedure

The queen is shaved and prepped for sterile surgery. Both ventral midline and flank approaches have been described; the former is more common. Make an incision from just cranial to the pubis to the umbilicus. The abdominal musculature is stretched thin; use caution when incising the linea alba.

Exteriorize the gravid uterus and pack it off with saline-moistened laparatomy sponges (Fig. 65-1). Make an incision in a relatively avascular area of the uterus in one horn near the uterine body. All kittens can be milked from both horns to this single incision readily. If there are a large number of kittens, an incision may be made in each uterine horn. If a kitten is stuck in the pelvic canal, an incision should be made in the uterine body.

The kitten is removed, still encased within its amniotic sac, and handed off to waiting personnel for resuscitation (see Chapter 64). The placenta and other fetal tissues should be removed with the kitten if they come readily. If the placenta still is tightly adhered to the endometrium,

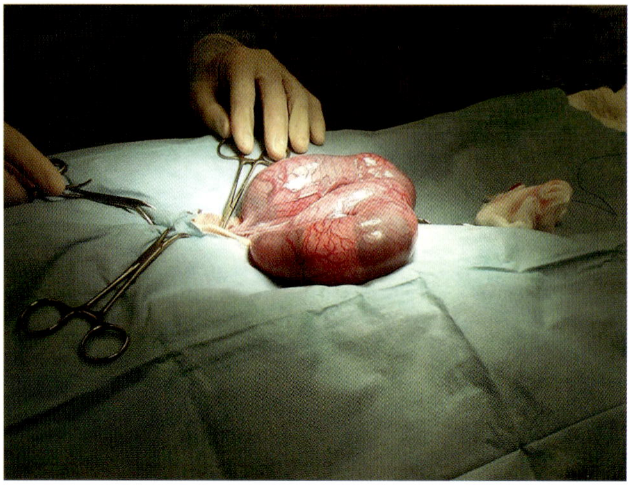

Figure 65-1: Exteriorized gravid cat uterus.

it may minimize bleeding if it is left in place and allowed to detach and pass with post-partum lochia.

When all kittens have been removed, palpate the entire uterus and into the pelvic canal to make sure no kittens have been missed. If the queen is not to be spayed, close the uterus with absorbable suture in an inverting pattern, in one or two layers. If the queen is to be spayed, the uterus is roughly closed in a continuous pattern to minimize leaking of uterine contents into the abdomen during manipulation, and routine OHE is performed. Flush and inspect the abdomen. Close the abdominal layers routinely. Non-absorbable suture may be used to ensure closure and healing of the thinned linea alba.

En-bloc OHE has been described as a means of performing concurrent OHE and C-section. The ovarian arteries and uterine arteries and body are clamped and the gravid uterus removed. The uterus is handed off to an assistant, who quickly opens the uterus and removes all the kittens for resuscitation. The surgeon returns to the queen to ligate the ovarian and uterine vessels and complete the OHE. The advantage of this technique is minimal transfer of anesthetic agents to the kittens and decreased possibility of abdominal contamination with uterine contents. The disadvantage is possibility of fetal stress secondary to loss of uterine and placental blood flow. There are few studies documenting success rates with this technique compared with standard C-section.

Post-care and complications

Little is known about pharmacokinetics of pain medications in lactating queens. Opioids enter milk, but total dose ingested by any given kitten is low, so this may be the preferred drug class for this indication.

Necessity of C-section to relieve dystocia at one parturition is not indicative of a requirement for C-sections at future queenings. For most queens that require more than one C-section it is because there is a common underlying cause for dystocia in both cases or because they are of a breed that is predisposed to dystocia.

Supplemental reading

Traas AM. 2008. Surgical management of canine and feline dystocia. *Theriogenology* 70:337–342.

Section VI

Feline reproductive physiology

66

What is the normal age for puberty onset in queens and toms?

General information

Puberty is defined in queens as first estrus and in males as acquisition of normal semen quality and ability to exhibit normal breeding behaviors. Puberty onset is dependent on age, body weight, breed, and season of the year.

Most queens first exhibit estrus between 4 and 12 months of age, when they reach a body weight of 2.3 to 2.5 kg (5.1 to 5.5 lb). Queens are seasonally polyestrous (see Chapter 67) and will not undergo puberal estrus when they reach an appropriate age or body weight if it is not also the correct time of the year. In the northern hemisphere, queens do not cycle when day length is decreasing (September through December), with that effect more marked as you move away from the equator. Breed also plays a role; in general, Burmese queens enter puberty earlier than other breeds, and long-haired breeds and Manx queens enter puberty later than other breeds.

Toms undergo puberty at 8 to 10 months of age on average, although there is evidence of spermatogenesis in many cats when younger. Season plays less of a role in puberty onset for male cats.

Clinical implications

Puberty onset in queens may be hastened by maintaining the queen in an environment with increased lighting, mimicking long day length. Conversely, cats housed in an environment with no exposure to natural light and minimal supplemental lighting may enter puberty later than expected.

Supplemental reading

Axner E. 2008. Updates on reproductive physiology, genital diseases, and artificial insemination in the domestic cat. *Reprod Dom Anim* 43:144–149.

Griffin B. 2001. Prolific cats: The estrous cycle. *Compendium* 23:1049–1056.

67

What are the normal parameters for the estrous (heat) cycle in queens?

General information

The heat or estrous cycle consists of five stages. These are proestrus, estrus, post-estrus, diestrus, and anestrus. Queens are polyestrous and seasonal, meaning they cycle continuously unless that cycle is broken by pregnancy, disease, or season. In the northern hemisphere, queens do not cycle when day length is decreasing, usually September through December. This effect is more marked as you move further away from the equator.

Proestrus is the stage where the queen is undergoing follicular development and starts to attract male cats. Queens, unlike bitches, do not exhibit vulvar swelling and exudation of vulvar discharge. Proestrus rarely is seen in queens; when the owner first notes that the queen is in heat, she usually is in estrus.

Estrus, or standing heat, is the stage where the queen allows the male to mount and breed her. Physical signs are minimal compared with those seen in the bitch. In one study of 187 estrous cycles, scant transparent vulvar discharge was observed 54.1% of the time. Behavioral signs of estrus in queens include monotonous yowling and assumption of the lordosis posture, where the forelimbs are pressed to the ground, the hindlimbs straightened, the back arched, and the tail held to one side (Fig. 67-1). Some queens also are more affectionate when in estrus. Queens bred three or more times during estrus usually are induced to ovulate. Length of estrus varies from 2 to 19 days, with an average duration of 6 to 8 days. Studies disagree as to whether induction of ovulation shortens duration of estrus in queens.

Post-estrus is the stage after an estrus during which the queen was not induced to ovulate. Follicles regress and new follicles emerge such that the queen is again in estrus in about 8 to 10 days.

Diestrus is the stage after ovulation has been induced, during which the queen is under the influence of progesterone. Successfully bred queens are pregnant during this stage. Queens that were induced to ovulate but are not pregnant will maintain the corpora lutea (CL) and progesterone production for an average of 40 days before lysing those CLs and returning to estrus.

Anestrus is the stage of reproductive quiescence. There are no specific physical or behavioral changes. Seasonal anestrus in queens lasts about 3 months, during the time of the year when day length is decreasing.

Figure 67-1: Queen in estrus exhibiting lordosis.

Clinical implications

Duration of stages of the estrous cycle may be irregular in queens aged less than 1 year or more than 10 years. Induction of ovulation and maintenance of CLs during diestrus can be used as an estrus-suppressing regimen in queens (see Chapter 72).

Supplemental reading

Griffin B. 2001. Prolific cats: The estrous cycle. *Compendium* 23:1049–1056.

Johnston SD, Root Kustritz MV, Olson PNS. 2001. *Canine and Feline Theriogenology*. Philadelphia: WB Saunders Co., pp. 396–405.

Pereira da Silva TF, Machado da Silva LD, Uchoa DC, et al. 2006. Sexual characteristics of domestic queens kept in a natural equatorial photoperiod. *Theriogenology* 66:1476–1481.

Root MV, Johnston SD, Olson PNS. 1995. Estrous length, pregnancy rate, gestation and parturition lengths, litter size, and juvenile mortality in the domestic cat. *J Amer Anim Hosp Assoc* 31:429–433.

68

What are the normal parameters for semen quality in cats?

General information

Semen collection and semen evaluation are described in Chapters 60 and 61, respectively. Semen evaluation in cats includes assessment of color, volume, percentage progressively motile spermatozoa, concentration and total number of spermatozoa in the ejaculate, and percentage morphologically normal spermatozoa (Table 68-1).

Clinical implications

Clear semen is indicative of lack of spermatozoa in the sample. Yellow color is urine contamination.

Volume varies with collection technique and is not a reflection of capabilities of the cat. Volume must be noted, however, as it is concentration multiplied by volume that yields total number of spermatozoa in the ejaculate.

Morphologic defects can be classified as primary (occurring during spermatogenesis) or secondary (occurring during storage in the epididymis or as an artifact of sample preparation). Examples of primary morphologic defects are double heads, double midpieces, double tails, bent midpieces, and proximal cytoplasmic droplets. Examples of secondary defects are bent tails, detached heads, and distal cytoplasmic droplets. Effect of a preponderance of any given defect on fertility is not well defined in cats. If most defects are primary, this suggests that a problem lies within the testis and prognosis for return to normal semen quality is worse than if most defects are secondary.

Table 68-1. Normal semen parameters in cats.

Parameter	Normal
Color	Moderately milky or opalescent
Volume	0.01 to 0.7 ml
Percentage progressively motile spermatozoa	70% or greater
Total number of spermatozoa in the ejaculate	3.5 to 60 million
Percentage morphologically normal spermatozoa	44 to 71%

Total number of spermatozoa in the ejaculate varies with the cat and with the collection technique (see Chapters 60 and 61).

Supplemental reading

Axner E, Linde-Forsberg C. 2007. Sperm morphology in the domestic cat and its relation with fertility: A retrospective study. *Reprod Dom Anim* 42:282–291.

Johnston SD, Root Kustritz MV, Olson PNS. 2001. *Canine and Feline Theriogenology*. Philadelphia: WB Saunders Co., pp. 508–520.

Section VII

Feline reproductive management

69

What age is best for ovariohysterectomy of female cats?

General information

Although studies have demonstrated safety of ovariohysterectomy (OHE), or spay, in cats as young as 7 weeks of age, there are no prospective long-term studies demonstrating the optimal age for OHE of female cats. Non-owned cats (stray or feral animals, humane society animals) should be spayed prior to placement in a new home. There is much in the popular literature on this topic; veterinarians are cautioned to be aware of the number and validity of studies cited to support statements of fact in any publication.

Advantages of OHE include decreased incidence of mammary neoplasia in aged cats; decreased incidence of uterine disease, most notably pyometra; and decreased incidence of sexually dimorphic behaviors.

Mammary neoplasia is the third most common tumor type in female cats. Greater than 90% of reported cases are malignant, with spread locally and to regional lymph nodes, lungs, and other tissues. Multiple studies over decades have documented a protective effect of OHE; it has been reported that sexually intact female cats have seven times the risk of developing a mammary tumor when aged compared with spayed cats. Surgery is the preferred treatment modality.

Pyometra is common in aged female cats; there is a significant likelihood that cats will have uterine disease by 5 years of age. OHE is curative of pyometra, with a reported mortality rate of 8%.

The most significant disadvantage of OHE in cats is obesity. Some studies suggest a predisposition to diabetes mellitus, especially in Burmese cats; this may be associated with obesity. Obesity is the most common nutritional disorder in cats, and multiple retrospective studies have demonstrated a correlation between spay or castration and weight gain. In cats, it has been demonstrated that gonadectomy is associated with decreased metabolic rate. Weight gain is easily controlled by the owner, with appropriate diet and exercise. Another reported concern is general surgical risk; adverse postsurgical events are reported to occur up to 33% of the time in queens, with most reported side effects mild and self-resolving.

Clinical implications

The veterinarian and client must take into account incidence and morbidity and mortality of disorders associated with OHE to determine the optimal age of OHE for a given cat. Breed also may play a role; Siamese and domestic Japanese breeds of cat are predisposed to mammary neoplasia. The high incidence and high morbidity and mortality associated with mammary neoplasia in cats argue strongly for OHE before the first estrus, which may occur at as young as 4 months of age (see Chapter 66).

Supplemental reading

Romagnoli S. 2008. Surgical gonadectomy in the bitch and queen: Should it be done and at what age? Proceedings, Southern European Veterinary Conference, Barcelona, Spain.

Root Kustritz MV. 2007. Determining the optimal age for gonadectomy of dogs and cats. *J Amer Vet Med Assoc* 231:1665–1675.

70

When is it best to perform ovariohysterectomy of female cats relative to heat?

General information

Vascularity of the ovaries and uterus is increased when cats are under the influence of estrogen, as in proestrus and estrus, and during pregnancy. Vascularity is decreased after ovulation, when the queen is under the influence of progesterone, unless the queen becomes pregnant. Most cats are best spayed between heat cycles, when in post-estrus or anestrus (see Chapter 67).

Cats that are induced to ovulate may develop mammary hypertrophy during diestrus, or non-neoplastic enlargement of one or more mammary glands (Fig. 70-1). This condition also may occur in animals receiving exogenous progestins.

Clinical implications

Because of the increased vascularity, anesthesia and surgery take longer, and there is a greater risk of intra- and postoperative bleeding if surgery is performed during proestrus or estrus.

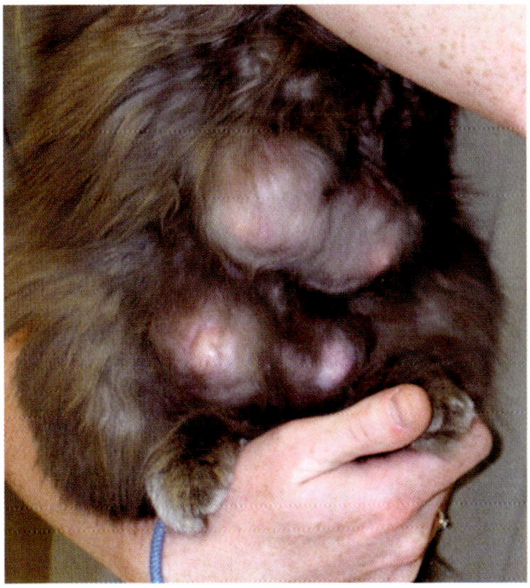

Figure 70-1: Mammary hypertrophy in a queen.

FELINE REPRODUCTION: II

Similar risks, along with dehydration and anemia, are present if OHE is performed during pregnancy, as the heavily gravid uterus is removed surgically. There are no published studies documenting increased incidence of side effects relative to the stage of pregnancy during which OHE is performed.

OHE is curative for mammary hypertrophy, which will regress after the progesterone stimulus is removed. Surgery may be postponed until mammary glands regress in size, or a flank approach may be necessary in severe cases.

Supplemental reading

Johnston SD, Root Kustritz MV, Olson PNS. 2001. *Canine and Feline Theriogenology*. Philadelphia: WB Saunders Co., pp. 474–477.

71

What age is best for castration of male cats?

General information

Although studies have demonstrated safety of castration in cats as young as 6 weeks of age, there are no prospective long-term studies demonstrating the optimal age for castration of male cats. Non-owned cats (stray or feral animals, humane society animals) should be castrated prior to placement in a new home. There is much in the popular literature on this topic; veterinarians are cautioned to be aware of the number and validity of studies cited to support statements of fact in any publication.

The advantages of castration include decreased incidence of sexually dimorphic behaviors and decreased incidence of diseases of the testes. The sexually dimorphic behaviors best controlled by castration are aggression and urine marking.

The only significant disadvantage of castration is weight gain. Obesity is the most common nutritional disorder of cats, and multiple retrospective studies have demonstrated a correlation between castration and weight gain. In cats, it has been demonstrated that gonadectomy is associated with decreased metabolic rate. Weight gain is easily controlled by the owner with appropriate diet and exercise.

The popular literature and tradition cite increased incidence of urethral blockage as a complication of castration in male cats. Multiple studies have demonstrated that there is no correlation between castration at any age and incidence of urethral obstruction in cats. One study has demonstrated a possible increase in incidence of feline lower urinary tract disease, defined as hematuria in the absence of urinary tract infection, in castrated animals.

Clinical implications

The normal behavior of the intact male cat makes him a poor candidate for a house pet and, in many cases, a danger to other animals and to people in the home. The author strongly recommends castration well before puberty of male cats not intended for breeding (see Chapter 66).

Supplemental reading

Root Kustritz MV. 2007. Determining the optimal age for gonadectomy of dogs and cats. *J Amer Vet Med Assoc* 231:1665–1675.

72

Are nonsurgical alternatives available for sterilization or contraception of female or male cats?

General information

Contraception is reversible control of reproduction; sterilization is nonreversible or permanent control of reproduction. Most often we are interested in the latter in companion animal medicine. As of this writing, there are no nonsurgical sterilants commercially available that are completely effective and safe for use in dogs and cats. This discussion will primarily concern nonsurgical contraceptives.

Females

Hormones

Progesterone

For female dogs, a synthetic form of progesterone, megestrol acetate (Ovaban™), was commercially available for many years and is still an approved estrus-suppressing drug for breeding and nonbreeding bitches. It is not available commercially at this time but can be dispensed as a human brand of megestrol acetate through a human pharmacy. Megestrol acetate works by exerting negative feedback to the pituitary, preventing release of follicle-stimulating hormone and luteinizing hormone. Megestrol acetate has never been approved for use in cats, although a regimen of 5 mg/cat per os for 5 days and then weekly has been published. Side effects include increased appetite, mammary development, uterine disease, diabetes mellitus, and possibly mammary neoplasia.

Many other forms of progesterone are available as contraceptives in other countries. These include injectable forms of progesterone, such as medroxyprogesterone acetate, and unique progestin compounds reported to have minimal side effects, such as proligestone. In many countries where performance of elective ovariohysterectomy (OHE) and castration are unethical or illegal, progestins routinely are administered as injections or via a patch, implant, or other slow-release systems. These compounds are not available in the United States.

Androgen

A mild androgen, mibolerone (Cheque™), is approved for use in nonbreeding bitches in the United States. This drug is not approved for use in breeding bitches and has never been approved for use in cats. Cheque™ is no longer available commercially, but mibolerone can be procured through compounding pharmacies. Androgens inhibit estrus by exerting negative feedback on the pituitary. In cats, the effective dose of mibolerone for estrus suppression is nearly the toxic dose. Side effects of therapy include cervical skin thickening, clitoral hypertrophy, and, if toxicity occurs, hepatic failure and death. Mibolerone is not recommended for use in cats.

Ovulation induction

Queens that are induced to ovulate but do not become pregnant will maintain high progesterone for an average of 45 days after ovulation induction. Although this is not a long-term solution, intermittent ovulation induction will permit queens to be shown or transported when not in heat. Ovulation may be induced with physical manipulation of the vagina, but ovulation usually is induced with gonadotropin-releasing hormone (GnRH: 25 μg/cat intramuscularly). It is not known whether recurrent ovulation induction and subsequent prolonged exposure to progesterone may predispose treated queens to pyometra.

Immunocontraception

Immunocontraception relies on the body's immune response to block fertility by stimulating the creation of antibodies against specific molecules or tissues. The two antigens best described are zona pellucida (ZP) proteins, which encircle the egg, and GnRH.

ZP proteins are highly conserved between species. This permits use of readily available pig ovaries as a source of antigen. At present, there is no commercially available ZP vaccine for cats. Research has documented an antibody response in queens immunized against ZP proteins from many species, but these antibodies do not prevent conception.

Males

Very little work has been done to demonstrate efficacy of contraceptives in male cats. Because the normal sexual behavior of male cats is so aggressive, most people prefer castration as a means of controlling reproduction and reproductive behaviors.

Clinical implications

Most pet owners are as concerned about controlling their queen's or tom cat's physical and behavioral manifestation of reproduction as they are in controlling reproduction itself. Current nonsurgical methods either are not efficacious in controlling reproduction or permit continuing physical changes of heat and reproductive behaviors that owners consider objectionable. Surgery (OHE or castration) is the standard for reproduction control in the United States at this time (see Chapters 69 to 71).

Supplemental reading

Kutzler M, Wood A. 2006. Non-surgical methods of contraception and sterilization. *Theriogenology* 66:514–525.

Levy JK, Mansour M, Crawford PC, et al. 2005. Survey of zona pellucida antigens for immunocontraception of cats. *Theriogenology* 63:1334–1341.

73

What comprises recommended pre-breeding evaluation for queens?

General information

A complete physical examination should be performed. Vaccinations should be current and the animal should be tested or treated for internal and external parasites. Feline leukemia testing should be performed on all group-housed and breeding animals, and feline leukemia vaccination should be discussed. Some advocate testing of all cats for feline coronavirus and only housing and breeding positive animals with positive animals, to prevent the spread of this ubiquitous virus and potentially decrease incidence of feline infectious peritonitis (FIP). Because heritability of susceptibility to FIP may be high, some advocate removing positive cats from breeding programs.

Cats have two blood type antigens, A and B. Type A cats have natural weak antibodies to the type B antigen; type B cats have natural strong antibodies to the type A antigen. If a type B queen is bred to a type A tom, she may produce type A kittens who will undergo hemolytic crisis when they ingest her colostrum, which will contain her strong anti-A antibodies and destroy their red blood cells (see Chapter 99). For this reason, all breeding cats should be blood typed before breeding and like bred to like. Blood types vary by breed; Persians and some British breeds have greater prevalence of type B cats in their population than other breeds.

Clinical implications

Much of pre-breeding evaluation for cats is preventative, promoting wiser breed pairings and housing of animals in the cattery.

Supplemental reading

Bell ET, Malik R, Norris JM. 2006. The relationship between the feline coronavirus antibody titre and the age, breed, gender and health status of Australian cats. *Aust Vet J* 84:2–7.

Bucheler J, Giger U. 1993. Alloantibodies against A and B blood types in cats. *Vet Immunol Immunopath* 38:283–295.

Johnston SD, Root Kustritz MV, Olson PNS. 2001. *Canine and Feline Theriogenology*. Philadelphia: WB Saunders Co., p. 406.

74

How do I use progesterone and other measures for breeding management?

General information

Because queens are induced ovulators, breeding management is less concerned with identifying significant events in the cycle and more concerned with providing optimum conditions for conception to occur.

Most queens exhibit overt behaviors of heat. Some queens may be more subtle in their estrous behavior, making estrus onset difficult to detect. Vaginal cytology can be used to identify estrus in these queens (see Chapter 58). Queens with cornified vaginal cytology should be presented to the male, and if breeding does not occur, artificial insemination can be considered.

The queen usually is brought to the tom, who is more likely to exhibit normal breeding behavior in his home territory. The male will grasp the scruff of the queen's neck in his teeth to pull her into position, introduce his penis, ejaculate, and release the queen within seconds. The queen will cry out as intromission of the penis and ejaculation occur (the coital cry) and, after the male releases her, will throw herself frantically about, rolling and licking at her vulva (the after-reaction). The coital cry and after-reaction may be taken as evidence that breeding occurred. At some point after the after-reaction is finished, the male will try to approach the queen again and she may or may not permit breeding.

The most common trigger for ovulation induction in queens is coitus. It has been well demonstrated that queens must be bred an average of three to four times during a given estrus for ovulation induction to occur reliably.

Clinical implications

Breeders often put cats together when they perceive the female to be in heat but do not observe or quantitate the breedings. It is valuable for the person responsible for the breedings to observe, both to make sure breeding did take place and to count the number of time coitus occurred. Some queens will allow breeding four times in one afternoon, while other queens will have to be introduced to the male over several consecutive days. If the owner is unsure whether or not breeding occurred or if breeding happens with sufficient frequency without subsequent conception and pregnancy, measurement of serum progesterone can be used to determine if ovulation occurred. Serum progesterone should be greater than 5 ng/ml by 1 week after the last breeding. If serum progesterone has not risen, medical induction of ovulation may be attempted (see Chapter 86).

Supplemental reading

Concannon PW, Hodgson B, Lein D. 1980. Reflex LH release in estrous cats following single and multiple copulations. *Biol Reprod* 23:111–117.

Johnston SD, Root Kustritz MV, Olson PNS. 2001. *Canine and Feline Theriogenology*. Philadelphia: WB Saunders Co., pp. 396–413.

Root MV, Johnston SD, Olson PNS. 1995. Estrous length, pregnancy rate, gestation and parturition lengths, litter size, and juvenile mortality in the domestic cat. *J Amer Anim Hosp Assoc* 31:429–433.

75

What drugs are unsafe to use during pregnancy in queens?

General information

Drugs can be classified using the following designation: A = probably safe in cats, based on the lack of toxicity or teratogenicity identified in other species; B = safe if used cautiously; C = potentially risky and to be used only when the benefit clearly outweighs the risks; and D = contraindicated in all cases. Few drug studies have been completed in queens, requiring extrapolation from work in other species, most commonly humans and laboratory animals (Table 75-1).

Clinical implications

All drug use during pregnancy requires careful weighing of benefits and detriments. Because of the lack of research in queens and the subsequent need to extrapolate from other literature, good client communication regarding extra-label use of drugs is critical.

Table 75-1. Safety of drugs for use in pregnancy.

Chemical Name	Safety Designation	Comments
Antibiotics		
Ampicillin	A	
Amoxicillin	A	
Amoxicillin-clavulanate	A	
Cephalosporins	A	
Chloramphenicol	C	May decrease protein synthesis in fetus.
Doxycycline	D	All tetracyclines may cause bone and teeth malformations in fetus.
Erythromycin	A	
Fluoroquinolones	C	All drugs in this class have been associated with cartilage damage.
Gentamicin	C	Aminoglycosides are associated with ototoxicity and nephrotoxicity.
Metronidazole	C	Teratogenic in laboratory animals.
Trimethoprim–sulfa	B	

Table 75-1. *Continued*

Chemical Name	Safety Designation	Comments
Preanesthetics, anesthetics, sedatives		
Acepromazine	B	May cause fetal central nervous system depression.
Atropine	B	May cause fetal tachycardia.
Diazepam	C	
Glycopyrrolate	B	
Isoflurane	B	
Ketamine	B	May induce premature labor.
Naloxone	A	
Oxymorphone	B	
Sevoflurane	B	
Thiopental	C	May cause respiratory depression in fetus.
Parasiticides		
Diethylcarbamazine	A	
Fenbendazole	A	
Ivermectin	A	
Milbemycin oxime	A	
Piperazine	A	
Praziquantel	A	
Pyrantel	A	
Selamectin	A	
Thiabendazole	B	High doses cause toxemia in ewes.
Pain medications		
Acetaminophen	D	Unsafe for use in cats at any time.
Aspirin	D	Unsafe for use in cats at any time.
Carprofen	C	Increased gestation length and increase in stillbirths identified in laboratory animals.
Ibuprofen	D	Unsafe for use in cats at any time.
Selective COX-2 inhibitors	D	In humans, ingestion of drugs of this class may retard renal development in neonates.
Endocrine steroids		
Dexamethasone	C	Increased incidence of cleft palate and other congenital defects; may induce premature labor. High doses at mid-gestation may cause pregnancy termination.
Diethylstilbestrol	D	May induce abnormal development of reproductive system.
Prednisolone	C	

76

How can I terminate pregnancy in queens?

General information

Pregnancy maintenance in all species requires continuous secretion of progesterone. The source of gestational progesterone varies by species. In queens, the source of gestational progesterone is not well defined, with some studies indicating the corpora lutea (CL) as the sole source and others identifying both luteal and placental production. Drugs that lyse the CL are not as effective as pregnancy-terminating agents in queens as in bitches. No drugs are approved for pregnancy termination in queens in the United States (Table 76-1).

Clinical implications

Knowledge of the stage of pregnancy best permits the veterinarian to tell the owner what to expect; for this reason, the author prefers ultrasound as a pregnancy diagnostic technique. If pregnancy is terminated before 40 days, virtually all evidence of pregnancy will be resorbed. From 40 to 50 days, fluid and tissue will be passed through the vulva. After 50 days, recognizable fetal tissues will be passed. After 55 days, kittens may be born alive and should be euthanized upon passage. No one has demonstrated detriment to the queen with pregnancy termination later in gestation, but most veterinarians and owners will not perform termination late in pregnancy for aesthetic reasons.

Supplemental reading

Baldwin C, Evans LF, Peter AT. 2000. Evaluation of natural prostaglandin therapy for pregnancy termination in the domestic cat. *Fel Prac* 28:16–21.

Eilts BE. 2002. Pregnancy termination in the bitch and queen. *Clin Tech Sm Anim Prac* 17:116–123.

Fieni F, Martal J, Marnet PG et al. 2006. Clinical, biological and hormonal study of mid-pregnancy termination in cats with aglepristone. *Theriogenology* 66:1721–1728.

Georgiev P, Wehrend A. 2006. Mid-gestation pregnancy termination by the progesterone antagonist aglepristone in queens. *Theriogenology* 65:1401–1406.

Table 76-1. Pregnancy termination protocols reported for the queen.

Drug	Regimen	Side effects	Notes
ESTROGENS—suppress movement of eggs, affect implantation			
Estradiol cypionate ("Mismate shot")	0.125 to 0.25 mg intramuscularly (IM) 2 to 3 days after coitus intramuscular one time	1. Bone marrow toxicity—aplastic anemia, thrombocytopenia, leukopenia 2. Pyometra (see Chapter 80)	Behavioral estrus will be extended in duration.
PROSTAGLANDINS—lyse corpora lutea (CL); decrease progesterone; cause uterine contractions			
Prostaglandin F2alpha (Lutalyse™)	0.5 to 1.0 mg/kg subcutaneously given twice 24 h apart after day 40 of pregnancy or 2 mg IM once daily for 5 consecutive days after day 30 of pregnancy	1. Ptyalism, vomiting, diarrhea	Side effects include panting and mydriasis, and usually occur within minutes of drug administration, subsiding within 1 to 3 h.
PROLACTIN INHIBITORS—decrease function of CL (decrease progesterone)			
Cabergoline (Dostinex™)	5–15 μg/kg once daily per os for 5 days after mid-gestation	—	
PROGESTERONE RECEPTOR BLOCKERS			
Aglepristone (Alizine™)	10 mg/kg or 15 mg/kg subcutaneously (SQ) once daily for 2 consecutive days after mid-gestation		87% success rate reported. Depression and anorexia were exhibited briefly at the time of fetal expulsion in 9% of queens. Not available in the United States.

77

Should queens be spayed at the time of removal of a mammary mass?

General information

Mammary neoplasia is the third most common tumor type in female cats. Malignant adenocarcinoma is virtually always the tumor type present. In humans, genetics and estrogen are causative factors. In queens, Siamese cats are considered to be at high risk, but whether this is due to a genetic predisposition or to a longer lifespan in that breed has not been defined. Similarly, not all feline mammary tumors contain estrogen receptors. However, early ovariohysterectomy (OHE) is associated with decreased incidence of mammary neoplasia in aged female cats. The protective effect of OHE is greatest if surgery is performed before the queen goes through estrus, with a decreasing benefit with each heat cycle.

Surgery is the treatment of choice for mammary neoplasia. Concurrent OHE often is recommended. The arguments are that decreased exposure to estrogen and progesterone with subsequent heat cycles may prevent recurrence of neoplasia, and that performing both surgeries at once minimizes the total number of anesthetic episodes in the animal's life. The author is unaware of published studies demonstrating any beneficial effect of OHE at the time of surgical removal of mammary tumors.

Clinical implications

There is not overwhelming evidence to support concurrent OHE and mammary tumor removal as a method of decreasing mammary tumor recurrence and prolonging survival time in queens. However, incidence of other conditions of aged queens, specifically pyometra, will be decreased by OHE. The greatest benefit may be minimizing the number of anesthetic episodes for a given queen. If OHE and mammary tumor removal are performed at the same time, the OHE should be performed first to prevent seeding of the abdomen with tumor cells.

Supplemental reading

Johnston SD, Root Kustritz MV, Olson PNS. 2001. *Canine and Feline Theriogenology*. Philadelphia: WB Saunders Co., pp. 474–485.

Section VIII

Feline disease

78

How do I diagnose and treat dystocia?

Etiology

Incidence of dystocia, or abnormal queening, is reported as 3 to 6%. The risk of dystocia is increased in brachycephalic breeds and in some doliocephalic breeds.

Dystocia is best defined by comparison with normal queening. Normal queening occurs in three stages. Stage I of labor is the long stage during which the cervix dilates. The queen acts restless, may show nesting behavior, pants, is often inappetent, and may vomit. Stage II is active contractions and passage of the fetuses. A small volume of clear, tan, or slightly blood-tinged fluid may be passed before any kittens are born. Once active contractions are seen, a kitten should be born within 4 h; if the queen is pushing hard constantly, a kitten should be passed within 30 min. No more than 2 h should be allowed to pass between kittens. Kittens may be passed within the amniotic sac or the sac may rupture as the kitten passes through the birth canal. At birth, the queen should tear open the sac, if present, and vigorously lick the kitten to stimulate respiration. Stage III is passage of the placentas. Stages II and III often alternate. Placentas should be passed within about 15 min of passage of a kitten. The normal feline placenta is olive green to greenish-brown in color due to pigmentation in the marginal hematomas on the edge of the placenta. The queen may try to eat the placentas; this is an atavistic behavior that serves to keep the den clean to avoid attraction of predators, and the placentas have no nutritional or therapeutic value. Queening usually occurs at night and may be prolonged compared with whelping in the bitch; queens have been reported to give birth to live kittens over several calendar days.

Maternal causes of dystocia include uterine inertia, inadequate size of the birth canal, and abnormality of the pregnancy. Uterine inertia can be primary (no propulsive uterine contractions) or secondary (muscle fatigue after unproductive uterine contractions). Primary uterine inertia may be associated with hypocalcemia and may have a hereditary component. Primary uterine inertia is the most common reported cause of dystocia in queens. Secondary uterine inertia usually occurs in the presence of obstruction preventing passage of a kitten into the birth canal. Inadequate size of the birth canal may occur in very small queens. An example of abnormality of pregnancy is uterine torsion, which is uncommon.

Fetal causes of dystocia include fetal oversize and abnormalities of presentation, posture, and position. Fetal oversize may be absolute (normal birth canal, oversize kitten) as in small litters, or relative (part of kitten too big to fit through normal birth canal) as in kittens with hydrocephalus. Presentation is defined by what part of the kitten first enters the birth canal. Cranial

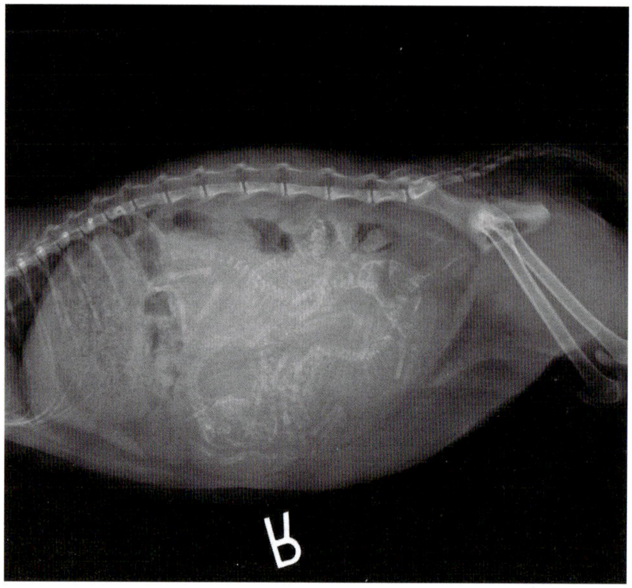

Figure 78-1: Lateral abdominal radiograph from a queen at the time of parturition.

presentation (muzzle and extended forelimbs presenting to the birth canal) and caudal presentation (tail and extended rear limbs presenting) are normal. Examples of abnormal presentations include breech (tail and flexed limbs presenting) and transverse (spine presenting). Position is defined by disposition of the extremities relative to the body. An example of abnormal position is neck flexed to the side.

Brachycephalic breeds of cat are at increased risk of dystocia compared with mesocephalic breeds. Other disorders associated with dystocia are prolonged gestation (more than 71 days from the last breeding), and abnormal vulvar discharge (green discharge prior to the birth of any kittens is indicative of placental separation).

Clinical signs

Queens in labor may look tired but should not appear systemically ill. Disorientation, vomiting, and aggression all are abnormal during active labor in queens. Normal vulvar discharge may be clear, tan, or slightly blood-tinged. It should not be purulent or frankly hemorrhagic, and should not be green if no kittens have been passed.

Diagnosis

History findings suggestive of dystocia are prolonged time limits (longer than 24 h of Stage I labor; more than 4 h since onset of Stage II labor or more than 30 min since onset of active, repetitive contractions; more than 2 h between kittens), overtly abnormal presentation of a kitten, or appearance of systemic illness, or abnormal vulvar discharge from the queen.

A brief but complete physical examination should be performed. Radiographs permit assessment of the number and size of kittens remaining in the uterus (Fig. 78-1). Viability cannot be determined accurately from radiography alone; signs of fetal death (gas within and around the fetus, collapse of the skull and axial skeleton) may not appear for up to 24 h after fetal death. Abdominal ultrasound is best used to determine viability by assessment of fetal heart rate.

Normal fetal heart rate should be about twice that of the queen. Fetal heart rate less than 170 bpm is indicative of fetal distress.

Treatment

Queens that appear healthy, have no evidence of fetal death or distress, and no apparent cause for dystocia may benefit from being left alone in a dark place with minimal observation for a time before any other measures are undertaken. Most queens give birth at night unattended. Overzealous observation and assistance from owners or veterinary professionals may slow parturition.

Manipulation of kittens with abnormal presentation or position is difficult in queens because of size restrictions. It is easy to dislocate joints and pull skin off of kittens while attempting manipulation. Instruments should not be used.

Oxytocin therapy is appropriate if radiographs have demonstrated normal size and likelihood that the kittens can pass through the birth canal. Doses of 0.5 to 2 IU can be given intramuscularly at 20- to 30-min intervals for no more than three doses. Repeated doses do not promote continuing uterine contractions because the uterus becomes refractory to oxytocin therapy as receptors fill on the myometrium and do not readily dissociate. Oxytocin increases frequency of uterine contractions; calcium increases strength of uterine contractions.

Queens carrying kittens that are in distress, that are carrying kittens too large to pass through the birth canal, or that are non-responsive to oxytocin therapy should undergo Cesarean section (C-section)(see Chapters 64 and 65). In one study of 1056 births, 8% resulted in C-section.

Supplemental reading

Gunn-Moore DA, Thursfield MV. 1995. Feline dystocia: Prevalence, and association with cranial conformation and breed. *Vet Rec* 136:350–353.

Root MV, Johnston SD, Olson PN. 1995. Estrous length, pregnancy rate, gestation and parturition lengths, litter size, and juvenile mortality in the domestic cat. *J Amer Anim Hosp Assoc* 31:429–433.

Sparkes AH, Rogers K, Henley WE, et al. 2006. A questionnaire-based study of gestation, parturition and neonatal mortality in pedigree breeding cats in the UK. *J Fel Med Surg* 8:145–157.

79

What diagnostic tests can be performed to identify the cause of pregnancy loss?

Etiology

Pregnancy loss can occur as a result of infectious or noninfectious causes. Infectious causes include bacterial infection; viral infections including feline herpesvirus, feline immunodeficiency virus, feline infectious peritonitis virus, feline leukemia virus, and feline panleukopenia virus; and protozoal infection with *Toxoplasma gondii*. Noninfectious causes include endocrine imbalance, developmental defects, and uterine disease.

Clinical signs

Pregnancy loss is associated with vulvar discharge and may or may not be associated with signs of systemic disease. Oftentimes, pregnancy loss is the presenting complaint. Signs that may be seen with specific infections are described under diagnosis.

Diagnosis

When presented with a pregnant cat that appears to be losing that pregnancy based on the presence of vulvar discharge or systemic signs of disease, a complete physical examination, abdominal ultrasound, drawing of blood for measurement of progesterone concentration in serum and complete blood count, and collection of a sample of vulvar discharge for aerobic culture should be performed. A thorough history should be taken and records reviewed to determine status of the queen for infection with any of the common viral diseases. If fetal heart rate is less than 200 bpm, fetal distress is present and the fetuses should be closely monitored for viability. If fetal heart rate consistently falls to less than 150 to 170 bpm, fetal death is imminent and pregnancy loss should be allowed to continue or the kittens removed via hysterotomy or hysterectomy.

If fetal tissues or aborted kittens are available for diagnosis, tissues should be harvested immediately after death, if possible, and the tissues should not be frozen prior to examination. Fresh samples should be refrigerated and submitted to a diagnostic laboratory, and fixed samples should be submitted as well. Tissues that should be submitted include liver, kidney, adrenal, small and large intestines, lung, heart, thymus, brain, fetal membranes, and fetal stomach content or fetal fluids. Swabs from the queen's vagina should be submitted in transport media.

Infectious causes

Bacteria

Causative bacteria of pregnancy loss reportedly are those that are part of the normal flora of the feline vagina. Other possibilities include *Salmonella* sp. and, very rarely, *Brucella canis*. Many queens with bacterial causes of pregnancy loss present with vulvar discharge as the only clinical sign of disease. Samples to be collected for bacterial culture include vulvar discharge, passed fetal tissues, and stomach contents of aborted kittens, which contain amniotic fluid and therefore accurately reflect the intrauterine environment.

Viruses
Any of the common feline viral diseases can be associated with pregnancy loss, which is usually associated with fetal death *in utero*. Other clinical manifestation of viral disease may or may not be evident.

Protozoa
T. gondii is the most common protozoal infection associated with pregnancy loss in cats. Systemic signs of disease always are present and include depression, diarrhea, ocular and nasal discharge, and anorexia. Definitive diagnosis requires demonstration of cysts of the organism in fetal tissues.

Noninfectious causes
Hypoluteoidism
Progesterone is required throughout pregnancy in all species. In queens, it has not been established where all progesterone is produced during gestation. Some research suggests that there is a transition from primarily luteal progesterone production to primarily placental production at about 40 days of pregnancy. Because pregnancy loss is commonly seen at this stage of gestation, it has been hypothesized that inadequate progesterone may be a cause of pregnancy loss in queens.

Developmental abnormalities
Small litters may be due to DNA abnormalities reflected as developmental defects incompatible with life. Teratogenic drugs also may induce abnormalities that are incompatible with life; the latter is more commonly associated with loss of an entire litter, while the former more commonly is reflected in loss of individual kittens.

Uterine disease
Cystic endometrial hyperplasia (CEH) and chronic uterine infection may be associated with pregnancy loss. In queens with severe uterine changes, chemical abnormalities may alter the function of spermatozoa. CEH may alter the movement of the fertilized ova, and implantation or placentation of the embryos. Severe uterine abnormalities may be visible on ultrasound. Definitive diagnosis requires uterine biopsy.

Treatment
Bacterial infections are treated with appropriate antibiotics, based on culture and sensitivity and acknowledging danger of some drugs during pregnancy if live offspring are present (see Chapter 75). Treatment for viral or protozoal infections usually is not rewarding for preventing further pregnancy loss.

Supplemental reading
Givens MD, Marley MSD. 2008. Infectious causes of embryonic and fetal mortality. *Theriogenology* 70:270–285.

Schlafer DH. 2008. Canine and feline abortion diagnostics. *Theriogenology* 70:327–331.

80

How do I diagnose and treat pyometra in queens?

Etiology

Pyometra is a two-step process. The first pathologic change is cystic endometrial hyperplasia (CEH), a thickening of the uterine lining that occurs secondary to repeated estrous cycling. Although queens are induced ovulators, CEH does develop in many queens with age. Research suggests that stimuli other than coitus may, on occasion, induce ovulation, making even unbred intact queens susceptible. After ovulation, the presence of elevated concentrations of progesterone promote hyperreactivity of the endometrium and gradual cystic hyperplasia. Development of CEH is a continuum and it is not completely understood why some queens develop pyometra with minimal CEH while other queens with severe CEH do not develop pyometra.

The second pathologic change is infection. Infection invariably is due to an organism that is part of the normal vaginal flora. *Escherichia coli* is the most common isolate. Pyometra is most common in queens that have never been pregnant, suggesting that pregnancy may have some sort of protective effect, presumably at the level of the endometrium.

Renal disease is a common sequel to pyometra. Endotoxins released from the cell wall of Gram-negative bacteria inhibit normal renal tubular function. These changes are reversible if infection is controlled quickly enough.

Clinical signs

Pyometra is most common in older queens that had been in estrus within the previous 4 weeks. There is no history of recent estrus in up to 22% of cases. Clinical signs vary with cervical patency.

Queens presenting with open-cervix pyometra have yellow-green to pink or red-tinged, thick, odoriferous vulvar discharge. Other signs include depression, inappetence, and vomiting. Affected cats may be febrile or hypothermic. Polyuria/polydipsia is less common in queens than in bitches.

Queens presenting with closed-cervix pyometra do not have vulvar discharge. Systemic signs of disease are as with open-cervix pyometra (depression, anorexia, vomiting, hyperthermia or hypothermia) but usually are more severe. Abdominal distension is evident.

Diagnosis

Diagnosis of pyometra requires demonstration of excessive fluid in the uterus and either purulent vulvar discharge (open-cervix) or systemic response indicative of infection (closed-cervix).

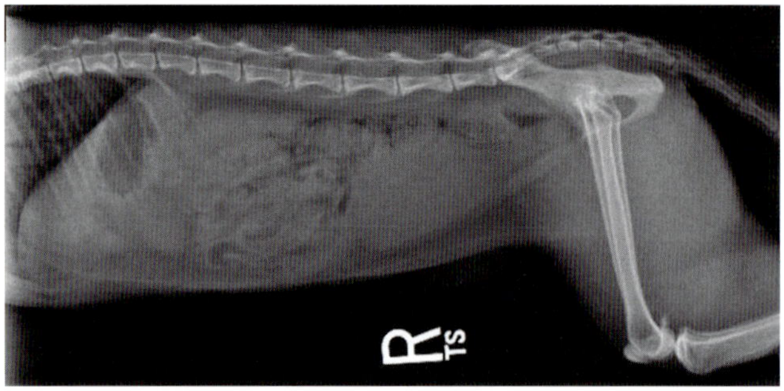

Figure 80-1: Lateral abdominal radiograph from a queen with pyometra.

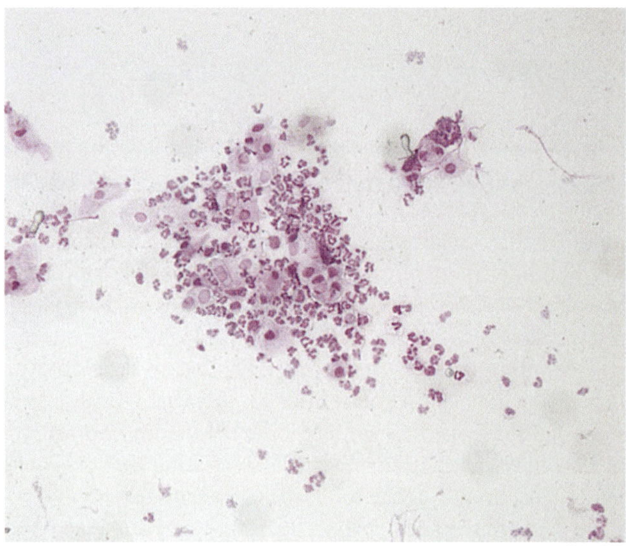

Figure 80-2: Vaginal cytology from a queen with pyometra.

Uterine enlargement can be identified by abdominal palpation; care should be taken in manipulating the friable uterus. Imaging is less directly traumatic to the uterus. Radiography can be used to identify uterine enlargement but cannot differentiate disease from pregnancy in animals that were bred (Fig. 80-1). Ulrasonography is preferred because it permits the operator to see within the uterus to identify free fluid. Pregnancy is readily identifiable using ultrasound by 16 to 25 days after breeding, permitting definitive differentiation from disease. Ultrasound cannot be used to determine if intrauterine fluid is purulent.

Cytology of vulvar exudate reveals full fields of degenerative polymorphonuclear cells (PMNs), bacteria, and non-cornified epithelial cells (Fig. 80-2). A sample of this discharge should be submitted for aerobic culture and sensitivity.

Blood should be drawn for a complete blood count and serum chemistry profile. The number of PMNs present may be very high, especially in animals with closed-cervix pyometra. Changes that may be evident on serum chemistry profile include azotemia and hyperproteinemia.

Table 80-1. Medical treatment of feline pyometra.

1. Assess uterine size in a repeatable manner (palpation, radiography, ultrasonography).
2. Collect a specimen of vulvar discharge for aerobic culture and sensitivity testing. Begin empirical therapy with ampicillin or amoxicillin. Change antibiotic if necessary based on culture and sensitivity results.
3. Administer prostaglandin F2alpha (0.25 mg/kg BID subcutaneously). Treat until uterine size nears normal (palpation, radiographs) or until no free intrauterine fluid is visible (ultrasound). Side effects of drug therapy include hypersalivation, panting, kneading at the abdomen, restlessness, and vocalization.
4. Send the queen home on antibiotics. Antibiotic therapy should continue for 1 month or until there has been no visible vulvar discharge for at least 1 week.
5. The queen is predisposed to pyometra at every subsequent cycle. Early in the next estrus, antibiotic therapy should be instituted. The queen should be bred at that first cycle after medical therapy for pyometra and should be spayed as soon as her breeding life is over.

Treatment

Ovariohysterectomy is the best treatment for pyometra in all cases. Surgical removal of the infected uterus immediately clears endotoxins from the body. CEH is not reversible so no medical therapy can return that queen to normal reproductive function.

Medical therapy may be considered if the queen meets the following criteria: (1) the cervix is open; (2) azotemia is absent or mild enough to be attributed to dehydration of other pre-renal causes; (3) the queen is still of breeding age; and (4) she is a valuable part of a breeding program. Medical therapy is not recommended in cases of closed-cervix pyometra. Treatment with drugs that cause or permit uterine contractions may help push purulent fluid against the internal cervical os with subsequent relaxation, but those drugs may also promote movement of fluid through the uterine tubes and into the abdomen, or cause uterine rupture, with subsequent peritonitis. Medical therapy with antibiotics and prostaglandin F2alpha is described here (Table 80-1).

Aglepristone, a progesterone receptor blocker, is available in other countries and is reported to be a successful component of medical therapy for pyometra in queens.

Supplemental reading

Agudelo CF. 2005. Cystic endometrial hyperplasia-pyometra complex in cats. A review. *Vet Q* 27:173–182.

Johnston SD, Root Kustritz MV, Olson PNS. 2001. *Canine and Feline Theriogenology*. Philadelphia: WB Saunders Co., pp. 463–471.

Nak D, Nak Y, Tuna B. 2009. Follow-up examinations after medical treatment of pyometra in cats with the progesterone antagonist aglepristone. *J Fel Med Surg* doi:10.1016/j.jfms.2008.09.006.

Schlafer DH, Gifford AT. 2008. Cystic endometrial hyperplasia, pseudo-placental endometrial hyperplasia, and other cystic conditions of the canine and feline uterus. *Theriogenology* 70:349–358.

81

How do I treat mastitis in a nursing queen?

Etiology

Mastitis is most commonly seen post-partum in the actively nursing queen. Environmental organisms and skin flora are those bacteria most commonly associated with the disease and include *Escherichia coli*, *Staphylococcus* sp. and *Streptococcus* sp.

Clinical signs

The infected gland is engorged, reddened, hot, and painful to the queen upon manipulation. More than one gland may be affected. The queen may be febrile with inappetence and lethargy, and may neglect the kittens. Localized abscesses may develop with gangrene of tissue. The very occasional queen will undergo septic shock.

Diagnosis

Mastitis is diagnosed by inspection. Cytology of fluid expressed from the mammary gland reveals bacteria and inflammatory cells. Expressed fluid from the affected gland(s) should be submitted for culture and sensitivity. Gangrenous tissue is black.

Treatment

Antibiotics that are likely to achieve therapeutic levels in the mammary gland are those that are lipid-soluble and are therefore also likely to be excreted into milk. Caution must be taken when using antibiotics with potential negative side effects on neonates (see Chapter 91). Kittens should be allowed to nurse unless gangrenous mastitis is present. The affected gland(s) should be covered to prevent excoriation of the friable tissue by kitten nails. A common empirical choice is amoxicillin–clavulanate. Abscessed areas should be incised, gangrenous tissue debrided, and drains placed.

Pain management in lactating queens is complicated by the ready movement of most analgesics into milk and the lack of research documenting potential effects on nursing kittens. Opioids pass readily into milk but can be reversed in neonates if signs of overdosing become apparent (lethargy, failure to nurse).

Supplemental reading

Johnston SD, Root Kustritz MV, Olson PNS. 2001. *Canine and Feline Theriogenology*. Philadelphia: WB Saunders Co., pp. 443–445.

Mathews KA. 2008. Pain management for the pregnant, lactating, and neonatal to pediatric cat and dog. *Vet Clin NA* 38:1291–1308.

82

What are the causes of persistent estrus (heat) in cats?

Etiology

The most common causes of true persistent estrus in queens are follicular ovarian cysts and ovarian tumors. Persistent estrus is caused by continuing secretion of estrogen. Estrogen production through the adrenal gland is possible but has not been reported as a pathologic phenomenon in queens. Exogenous estrogen administration, either as a "mismating shot" (see Chapter 76) or through dermal absorption of topical products applied to the owner, is another possible cause.

Clinical signs

Cats in estrus show behavior that many owners find annoying or disturbing. The client should be closely questioned to determine if the cat is truly in a persistent estrus or if that is a perception based on the severity of the behavior change and the normal frequent cycling of queens. True persistent estrus is defined as physical and behavioral signs of proestrus and estrus, and the presence of cornified vaginal cytology, for at least 4 weeks (see Chapter 67). Bilaterally symmetrical alopecia may be evident, especially if exposure to estrogen has been very prolonged. Abdominal distension and ascites may be evident in queens with granulosa cell tumor.

Diagnosis

Follicular ovarian cysts are visible by abdominal ultrasonography and usually are not large enough to distort the ovary. The most common ovarian tumor type that is functional, secreting estrogen, is granulosa cell tumor. Clinical signs may arise at any time, not necessarily at the expected time of estrus. Oftentimes, the ovary is so enlarged as to be palpable per abdomen as a cranial abdominal mass. Ultrasonographically, ovarian tumors are more likely to appear mottled and septate. Differentiation of follicular ovarian cysts and ovarian tumors also may be made by response to treatment. Similarly, diagnosis of exogenous estrogen as a source may be made by careful history taking and response to withdrawal of potential inciting agents. With any prolonged estrogen exposure, bone marrow suppression may occur, evidenced by nonregenerative anemia, leukopenia, and thrombocytopenia.

Treatment

Ovarian follicular cysts are treated by induction of ovulation, either with gonadotropin releasing hormone (Cystorelin), 25 μg/cat intramuscularly (IM) or human chorionic gonadotropin (500 IU

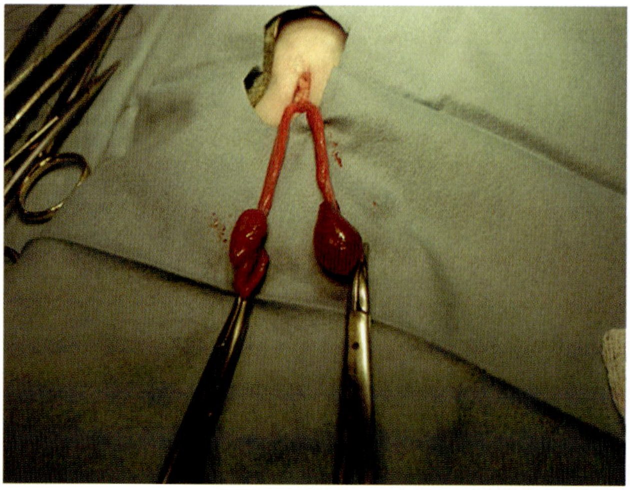

Figure 82-1: Gross appearance of follicular ovarian cysts in a queen.

IM), or by ovariohysterectomy (OHE) (Fig. 82-1). OHE may be the preferred therapy as cats induced to ovulate after persistent estrogen exposure may be predisposed to pyometra (see Chapter 80). Granulosa cell tumors are treated by OHE. Because metastasis and local invasion are uncommon, OHE usually is curative. Blood dyscrasias are treated with appropriate blood products.

Supplemental reading

Johnston SD, Root Kustritz MV, Olson PNS. 2001. *Canine and Feline Theriogenology*. Philadelphia: WB Saunders Co., pp. 456–462.

83

What are the causes of lack of cycling (persistent anestrus) in cats?

Etiology

Seasonal anestrus is normal in queens (see Chapter 67). Nonseasonal anestrus may be due to previous ovariohysterectomy (OHE), inadequate exposure to light during the breeding season, abnormalities of chromosome complement and sexual differentiation, and progesterone-secreting ovarian cysts or tumors. Malnutrition and poor body condition may be associated with lack of estrous activity.

Clinical signs

Regular observation of queens and intermittent exposure to toms may be required to identify behavioral estrus. Some queens housed in a colony will not demonstrate signs of estrus in the presence of a dominant queen; vaginal cytology may be required to identify follicular activity in these queens.

Diagnosis

Owners should be questioned regarding where the queen is housed and how they have tried to identify signs of heat. Other history questions may include diet, medications given, and whether there is any possibility of the queen having been spayed. Previous OHE may be identified by demonstration of a scar on the ventral midline or by demonstration of elevated concentrations of luteinizing hormone in the blood. Blood should be drawn for assay of progesterone in serum; progesterone concentration of greater than 2 ng/ml may indicate presence of a luteal ovarian cyst or progesterone-secreting tumor. Luteal cysts usually are not visible ultrasonographically; persistence of elevation in serum progesterone for more than 2 months is diagnostic. Chromosomal abnormalities can be identified by karyotyping, which is available at Texas A&M University (see Resources).

Treatment

If there is no clear evidence of systemic disease or another cause of lack of cycling, queens can be put on a good plane of nutrition and housed with adequate exposure to natural light or to 12–14 h of artificial light daily. Luteal ovarian cysts may be treated with administration of the luteolytic agent prostaglandin F2alpha. Side effects include panting, hypersalivation, vomiting, and defecation. These side effects usually subside within 3 h of treatment. Luteal cysts also can be treated by OHE.

Chromosomal abnormalities reportedly associated with persistent anestrus in queens include 37,XO, and cats with multiple cell lines. Some cats with an abnormal karyotype may demonstrate intermittent estrous cycling, but all are considered infertile.

In the absence of any defined cause, medical induction of estrus may be attempted. There are no drugs approved for this purpose in queens in the United States. No drug described can accurately induce estrus on a given date or with complete consistency. The regimen reported is follicle-stimulating hormone, 2 mg once daily intramuscularly (IM) for 3 to 7 days or until estrus onset, followed by natural mating or induction of ovulation with human chorionic gonadotropin, 250 IU IM.

Supplemental reading

Kutzler MA. 2007. Estrus induction and synchronization in canids and felids. *Theriogenology* 68:354–374.
Lyle SK. 2007. Disorders of sexual development in the dog and cat. *Theriogenology* 68:338–343.

84

What are the causes of estrus after ovariohysterectomy of female cats?

Etiology

The primary rule-out in all cases is ovarian remnant syndrome. Adrenal production of estrogen to the extent of causing clinical signs of estrus has not been identified in cats. It has been well demonstrated that ovarian tissue can vascularize and regain function after transplantation. Incidence of ovarian remnant syndrome is not correlated with reason for the ovariohysterectomy (OHE), with physical condition of the cat at the time of surgery, or with level of experience of the veterinary surgeon.

Clinical signs

Spayed female cats will show behavioral signs of estrus with the periodicity of intact cats (see Chapter 67). Signs may first appear anywhere from 7 days to 10 years after the OHE. In areas where cats cycle seasonally, signs often first appear at the beginning of the breeding season.

Diagnosis

The cat should be seen when the owner perceives her to be showing signs of estrus. Vaginal cytology is used to identify estrogen production by presence of cornified cells (see Chapter 58). This is evidence of follicular tissue. Ovulation is induced by administration of gonadotropin releasing hormone (25 µg/cat intramuscularly). Serum progesterone is assayed 2 to 3 weeks later. If it is greater than 2 ng/ml, luteal tissue is present. This is definitive for diagnosis of ovarian remnant as ovary is the only tissue that can produce estrogen and that can be induced to secrete progesterone.

Treatment

Surgery is the preferred treatment. There are no estrus-suppressing drugs approved for use in cats in the United States, and all those available for dogs have severe side effects in cats. Cats should not be allowed to continue cycling as that predisposes them to mammary neoplasia (see Chapter 69). Surgery should be performed when there is active tissue on the remnant to make it easier to see (Fig. 84-1). This can be carried out either when the cat is showing signs of behavioral estrus and has cornified vaginal cytology, in which case the surgeon is looking for follicles on the ovarian remnant, or after induction of ovulation, in which case the surgeon is looking for luteal tissue on the ovarian remnant. The advantage of the latter is that the cat will bleed less

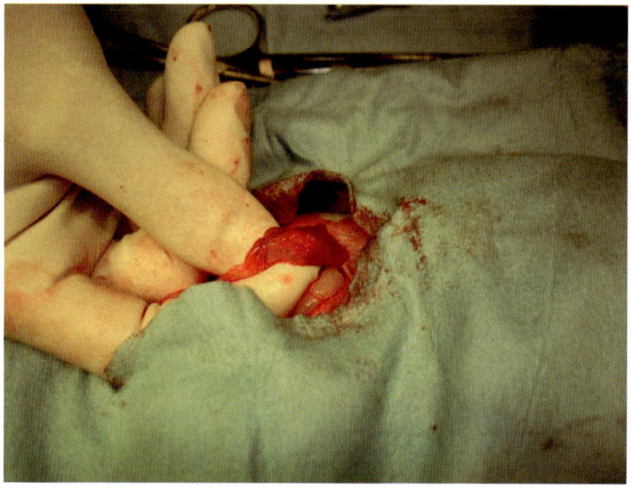

Figure 84-1: Ovarian remnant in a cat.

during surgery if she is not under the influence of estrogen and that luteal tissue persists longer than follicular tissue, increasing the odds that the tissue will still be evident when surgery is performed. Remnants may be present on both sides, and practitioners are strongly encouraged to explore both ovarian pedicles and remove any abnormal tissue. If there is no overtly abnormal tissue, remove the scar tissue in the area, taking care not to ligate the ureter. Rarely, ovarian remnants are found away from the ovarian pedicles. Any tissue removed should be submitted for histopathology; there are reports of neoplastic transformation of ovarian remnant tissue.

Supplemental reading

DeNardo GA, Becker K, Brown NO, Dobbins S. 2001. Ovarian remnant syndrome: Revascularization of free-floating ovarian tissue in the feline abdominal cavity. *J Amer Anim Hosp Assoc* 37:290–296.

Miller DM. 1995. Ovarian remnant syndrome in dogs and cats: 46 cases (1988–1992). *J Vet Diagn Invest* 7:572–574.

85

What are the causes of persistent male behavior after castration of male cats?

Etiology

Rule-outs for persistent male behavior after castration are retained testes and functional adrenal masses.

Clinical signs

Male behavior includes spraying of foul-smelling urine and aggression. These behaviors are often mediated by testosterone in cats, as evidenced by a significant decline after castration. Intact male cats often have thickened cervical skin and are more muscular than average.

Diagnosis

Testosterone secretion is readily evidenced by appearance of cornified spines on the cat's penis (Fig. 85-1). Measurement of testosterone in serum is problematic because it is secreted pulsatilely. If testosterone assay is preferred, blood should be drawn 60 min after intramuscular administration of gonadotropin-releasing hormone; a value of 3 ng/ml or greater is evidence of testosterone-secreting tissue.

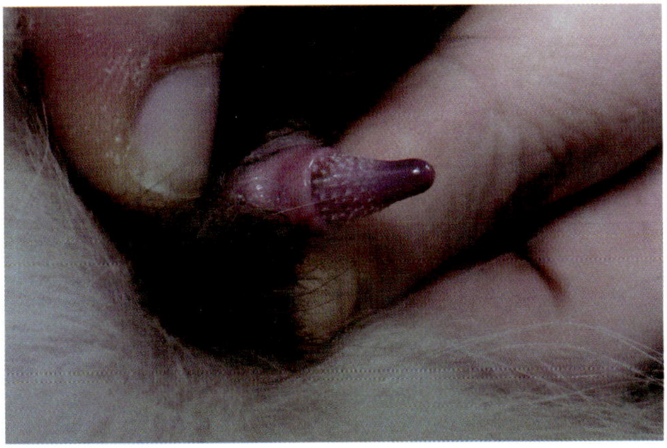

Figure 85-1: Cornified penile spines in a cat (photo courtesy of Dr. Anne Traas).

FELINE DISEASE

Treatment

Surgical exploration and removal of testosterone secreting tissue is required.

Supplemental reading

Millard RP, Pickens EH, Wells KL. 2009. Excessive production of sex hormone in a cat with an adrenocortical tumor. *J Amer Vet Med Assoc* 234:505–508.

86

What is the diagnostic approach for infertility of a queen?

Etiology

Causes of infertility in queens include improper breeding management, poor semen quality of the male (see Chapter 87), reproductive tract infection, systemic disease, non-receptive behavior due to pain or psychological factors, uterine pathology, anovulatory cycles, impatency of the reproductive tract, and advanced age of the queen. In one survey of seven queens presented for infertility, four had uterine pathology and three had no definitive cause for lack of conception or pregnancy loss.

Clinical signs

Queens with infertility may present for failure to stand to be bred, apparent conception failure, or known pregnancy loss. Because there is no early pregnancy test for cats (see Chapter 59), conception failure and early pregnancy loss cannot be differentiated. Queens with infection of the reproductive tract or uterine pathology often exhibit no specific clinical signs.

Diagnosis

See Table 86-1 for a diagnostic key. An accurate history must be taken of any past breeding attempts to determine if the queen was bred an appropriate number of times (see Chapter 74). Reproductive tract infection is best identified by culture of the vagina when the queen is in estrus. Behavioral causes are identified by observation of breeding.

Uterine pathology may or may not be visible ultrasonographically. Cystic endometrial hyperplasia may be visible as fluffy thickening of the endometrium and presumably interferes with fertility by altering the movement of spermatozoa from the cervix to the uterine tubes and interfering with normal implantation and placentation of conceptuses. Uterine pathology often is diagnosed at histopathology. Samples most commonly are retrieved at the time of ovariohysterectomy. Uterine biopsy requires laparotomy and hysterotomy and is rarely performed.

Impatency of the reproductive tract may be identifiable with hysterography, performed as vaginography when the queen is in estrus and the cervix open. More commonly, abnormalities of the tract are identified during laparotomy. Anovulatory cycles are defined as those cycles in which ovulation is not induced, evidenced as serum progesterone remaining below 2 ng/ml after three to four successful copulations.

Table 86-1. Diagnostic key for infertility of queens.

1a.	The queen is cycling.	2
1b.	The queen is not cycling.	3
2a.	The queen is induced to ovulate.	4
2b.	The queen is not induced to ovulate.	5
3.	Causes of lack of cycling are described in Chapter 83.	
4.	Causes of apparent lack of conception include poor semen quality, reproductive tract infection, uterine pathology, and impatency of the reproductive tract. Causes of pregnancy loss are described in Chapter 79.	
5.	Ovulation can be induced with gonadotropin-releasing hormone (25 μg intramuscularly) and mating attempted that day or the next, or artificial insemination performed.	

Treatment

Ensure the queen was bred successfully three to four times in estrus. Evaluate semen quality of the male if possible or ask for a record of success in recent breedings (see Chapter 61). Some queens and toms are not compatible genetically; if a queen has consistently failed to produce kittens with one male, recommend the use of males from different lines. Some breeds of cat are very recently derived from their wild cat ancestors; ensure the individuals being bred are not inbred or too closely related. Reproductive tract infection should be treated with appropriate antibiotics, based on culture and sensitivity.

Supplemental reading

Axner E, Agren E, Baverus V, et al. 2008. Infertility in the cycling queen: Seven cases. *J Fel Med Surg* 10:566–576.

Romagnoli S. 2003. Clinical approach to infertility in the queen. *J Fel Med Surg* 5:143–146.

Romagnoli S. 2005. Case report: Failure to conceive in the queen. *J Fel Med Surg* 7:59–63.

Root Kustritz MV. 2008. Infertility in a queen. *Vet Med* 6:23–24.

87

What is the diagnostic approach for infertility of a male cat?

Etiology

Infertility in male cats may be due to poor libido, inability to achieve intromission, and poor semen quality. Poor semen quality may be due to reproductive tract infection or to abnormal chromosome complement.

Clinical signs

Poor libido may be associated with systemic disease and evidenced by clinical signs associated with those conditions, for example, tachycardia and weight loss in aged toms with hyperthyroidism. Calico or tortoiseshell toms often are infertile as this coat color is associated with abnormal chromosome complement.

Diagnosis

Poor libido is diagnosed by observation of the tom with estrous queens. Inability to achieve intromission is difficult to observe and usually is inferred from lack of a coital cry or after-reaction by the queen (see Chapter 74). Lack of intromission also may be evidenced by lack of induction of ovulation, proven by lack of rise in serum progesterone to greater than 2 ng/ml after breeding. Penile abnormalities, either congenital or acquired (hair ring), are easily identified on physical examination. Semen collection and evaluation are described in Chapters 60 and 61. Seminal fluid culture may help identify reproductive tract infection. Chromosomal abnormalities can be identified by karyotyping, which is available at Texas A&M University (see Resources). In the absence of any identified cause of infertility, testicular biopsy may be performed as in male dogs (see Chapter 11).

Treatment

Hair rings can be removed with gentle traction and incision of the matted hair. Ovulation can be induced medically in the queen but obviously will not correct the infertility if ejaculation of semen of normal quality is not occurring. Infection can be treated with an appropriate antibiotic based on culture and sensitivity. Prostate disease is very rare in cats so practitioners need not be overly concerned about using antibiotics that penetrate the prostate.

Supplemental reading

Johnston SD, Root Kustritz MV, Olson PNS. 2001. *Canine and Feline Theriogenology*. Philadelphia: WB Saunders Co., pp. 544–548.

Section IX

Pediatric techniques

88

What are some of the techniques for venipuncture?

Anatomy

Blood cannot be drawn from the cephalic or saphenous veins of most pediatric animals because of their small size and the fragility of the vessels. The external jugular vein is the preferred site for venipuncture.

Pre-procedure considerations

The area over the vein should be moistened with water, not alcohol, to minimize heat loss. Use of a 25-gauge needle and tuberculin syringe minimizes collapse of the vessel or trauma to the vessel with subsequent hematoma formation.

Procedure

The animal may be restrained upright with forelimbs extended down and head extended, exposing the neck. Another technique that may be used is cradling the animal in dorsal recumbency in one's hand with the head near the wrist, holding down the forelimbs against the animal's sides with the pinky and thumb (Fig. 88-1). Restrict venous outflow at the thoracic inlet and draw blood as in adult animals.

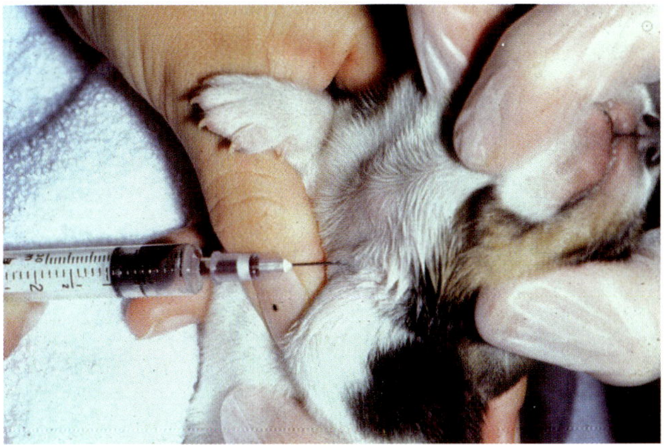

Figure 88-1: Restraint for venipuncture from a pediatric small animal (reprinted with permission from Root Kustritz MV. 2006. *The Dog Breeder's Guide to Successful Breeding and Health Management*. St. Louis, MO: Elsevier).

Post-procedure care and complications

Complications are as in adult animals and consist primarily of hematoma formation.

Supplemental reading

Root Kustritz MV. 2006. *The Dog Breeder's Guide to Successful Breeding and Health Management*. St. Louis MO: Elsevier, pp. 204–223.

89

What are some of the techniques for collection of a urine sample?

Anatomy

The urinary bladder of pediatric animals may be abdominal or pelvic, as in adults.

Pre-procedure considerations

The small size of the urinary bladder relative to the size of needles available for use of urine collection preclude cystocentesis as a safe method for urine collection in very small animals. There are no studies documenting safety of various urine collection techniques in pediatric dogs and cats.

Procedure

Within the first 3 weeks of life, puppies and kittens must be stimulated to urinate and defecate. Stimulate urination by manipulating the genitalia with a moistened cotton ball or cloth. As micturition occurs, urine can be collected with a syringe or allowed to drop into a sterile container, such as a red top tube. Another method involves use of "blue pads" or human incontinence pads, which have a porous surface and sterile cotton liner. Stimulate urination and allow the urine to fall onto the pad. Remove the outer surface layer and cut out the portion of the cotton layer that contains urine. Press this cotton layer in a syringe to remove the urine.

Post-procedure care and complications

There are no reported complications to urine collection as described above.

Supplemental reading

Root Kustritz MV. 2006. *The Dog Breeder's Guide to Successful Breeding and Health Management*. St. Louis, MO: Elsevier, pp. 204–223.

Valenti N. 2003. Another way to gather urine samples. *Vet Med* 98:304.

90

What is the technique for placement of an intraosseous catheter?

Anatomy

Intraosseous catheters can be placed in either the humerus or the femur. The author prefers placement in the femur, for ease of placement and maintenance of the catheter if it is to be indwelling (Fig. 90-1).

The head of the femur lies within the acetabulum. The greater trochanter of the femur is the ridge palpable at the hip joint as it is flexed and extended. Just medial to the greater trochanter, between the greater trochanter and the head, lies the trochanteric fossa.

Pre-procedure considerations

Intrasosseous catheter placement for administration of fluids, antibiotics, or any other substance that could be administered intravenously, is performed only in seriously ill neonatal animals. For this reason, anesthesia usually is not administered. Shave and surgically prepare the area over the trochanteric fossa.

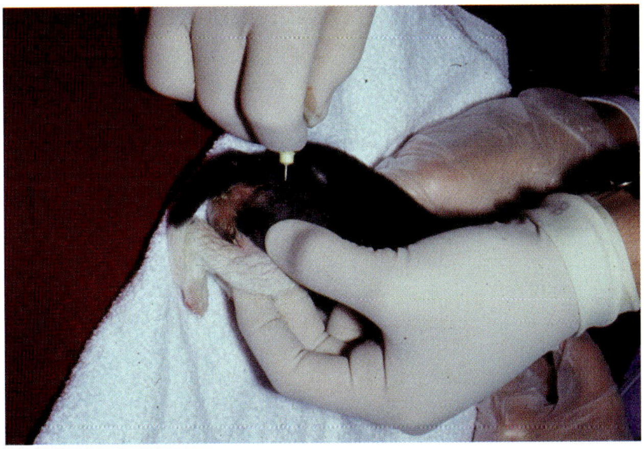

Figure 90-1: Placement of an intraosseous catheter in the femur (reprinted with permission from Root Kustritz MV. 2006. *The Dog Breeder's Guide to Successful Breeding and Health Management*. St. Louis, MO: Elsevier).

Procedure

Insert a 25-gauge 1-in. catheter into the trochanteric fossa, directing the stylette and needle along the marrow canal visually. The normal spiral configuration of the bone helps keep the needle within the marrow canal. Move the stifle and watch the hub of the catheter; if it is properly placed, movement of the stifle will cause synchronous movement of the catheter hub at the hip. You may be able to verify placement of the catheter by attaching a syringe and applying gentle suction. The dark blood of bone marrow will be evident if the catheter is in the marrow canal. You also can introduce a small volume of heparinized saline to check placement; fluid should flow freely. Bandage the catheter in place as you would in an adult animal. Maintenance fluid requirements for neonates are 60 to 200 ml/kg/day maintenance plus losses. Fluids and other substances may be administered using gravity or using a pump; the latter may better permit monitoring of adverse effects. Overhydration of neonates may cause cardiac overload, pulmonary edema, and intracranial hemorrhage, with subsequent decompensation and death.

Post-procedure care and complications

If the catheter is not placed properly, fluid will not flow freely or movement of fluid into the soft tissue surrounding the femur will be evident. It is not known whether disruption of the proximal femoral physis affects subsequent bone growth or whether disruption of the marrow cavity affects hematologic status in these pups. However, because the technique is used when the animal is near death and placement of intravenous catheters is difficult in small, dehydrated animals, placement of an intraosseous catheter is preferable to the alternative.

Supplemental reading

Freshman JL. 2005. Initially treating fading puppies and kittens. *Vet Med* 100:800–805.
Lawler DF. 2008. Neonatal and pediatric care of the puppy and kitten. *Theriogenology* 70:384–392.
Little S. 2006. How I treat orphaned kittens. *Waltham Focus* 16:2–6.

91

How do I determine safety and efficacy, and calculate correct dosage for drugs used in puppies and kittens?

Anatomy

Organs of distribution, metabolism, and excretion of drugs include the bloodstream, liver, lungs, kidneys, and intestinal tract.

Pre-procedure considerations

Small muscle size and reduced vascularity decrease the value of intramuscular injection as a route of drug administration for pediatric animals as absorption is slowed. Conversely, decreased subcutaneous fat in young animals compared with adults increases the rate of absorption of substances given subcutaneously. However, if the animal is hypothermic, blood flow to the subcutaneous space will be reduced and absorption slowed. The cardiovascular system of pediatric animals is unable to respond readily to increased volume by increased heart rate. Intravenous fluids and drugs must be administered carefully so as not to cause pulmonary edema. Vascularity of the extremities is relatively poor, especially in animals less than 3 weeks of age, who are not mobile and do not thermoregulate well. Hypothermia is associated with concentration of blood centrally and with decreased gastrointestinal (GI) motility. In general, the variable GI motility of young animals alters drug uptake. Renal function does not mature until 8 weeks of age; liver function does not mature until 5 months of age. Hepatic protein synthesis is less than in adults, decreasing the amount of protein available for binding to drugs in the bloodstream.

As a general rule, water-soluble drugs may not achieve therapeutic levels in plasma as the elevated relative percentage of body water permits the wide distribution of these drugs in the body. Lipid-soluble drugs may achieve very high concentrations in plasma as there is so little body fat in which they may be sequestered.

General principles for treatment of critically ill neonates include the following:

- For serious bacterial illness, avoid oral antibiotics during initial treatment because of the unpredictability of absorption of drugs through the GI tract. Subcutaneous, intravenous, or intraosseous routes are preferred.
- If the animal is near death, any antibiotic or other drug may be better than no therapy at all. Carefully weigh risks versus benefits.
- Every drug given carries some risk.

Table 91-1. Pharmacokinetics of commonly used drugs.

Drug	Route of Administration/ Dose	Tissue Where Metabolized	Route of Excretion	Comments
Amoxicillin/ ampicillin	PO/IM/SQ	Liver (partial)	Urine primarily, some excretion in feces	
Amoxicillin-clavulanate (Clavamox™)	PO/IM/SQ	Liver (partial)	Urine primarily, some excretion in feces	
Cephalosporins	PO/IM/SQ	Minimally metabolized, some hepatic conversion	Urine	IM injection may be painful.
Sulfadimethoxine (Albon™)	PO/IM/IV	Minimally metabolized in dogs, metabolized in liver in other species	Urine but reabsorbed by renal tubules so has long half-life	Highly protein bound.
Pyrantel pamoate	PO	Liver (partial)	Urine and feces	
Metronidazole	PO	Liver	Urine and feces	Absorption enhanced if given with food, lipid-soluble.
Praziquantel	PO/IM/SQ	Liver	Urine	Not recommended for pups less than 4 weeks of age or kittens less than 6 weeks of age.
Piperazine	PO	Kidney	Urine	
Fenbendazole	PO	Liver	Feces	
Thiopental	IV	Liver	Urine	Moderately protein-bound.
Oxymorphone	IV/IM/(SQ)	Liver	Urine	Metabolism prolonged in cats, lengthening drug half-life.
Propofol	IV	Liver	Urine	Highly lipid-soluble, highly protein bound.
Isoflurane/ sevoflurane	Inhalant	—	Exhaled	

PO = orally, IM = intrasmuscular, SQ = subcutaneous, IV = intravenous.

Procedure

There are very few publications documenting pharmacokinetics of commonly used drugs in puppies and kittens. Empirical dosing regimens include giving half the adult dose at the same frequency as in adults, and giving the adult dose less frequently. Veterinarians must take what they know about a given drug, and what they know about the animal's physiology, and use that information to decide how quickly a drug will be taken up, metabolized, and excreted (Table 91-1). For example, if you wished to treat a 4-week-old puppy with metronidazole, factors to consider would be that decreased hepatic function would lead to slower breakdown of the drug, decreased renal function would lead to slower excretion of the drug, and decreased body fat would lead to decreased distribution of the drug into body fat stores. All three of these would lead to increased levels in the bloodstream and prolonged half-life, suggesting that less frequent dosing and lower doses than those used in adults would be appropriate.

Antibiotics that in general should be avoided in puppies and kittens are chloramphenicol (blood dyscrasias), tetracyclines (enamel dysplasia and staining, inhibited or abnormal bone growth), potentiated sulfas (anemia, hepatitis), and aminoglycosides (ototoxicity, nephrotoxicity). Commonly used antibiotics and doses include amoxicillin (6 to 20 ml/kg twice daily per os), amoxicillin/clavulanate (12.5 to 25 mg/kg twice daily per os), and cephalexin (10 to 30 mg/kg twice daily per os).

Pain relief in pediatric animals is complicated by difficulty in identifying adverse side effects and by varying ability of young animals to metabolize and excrete medications in these classes. In general, nonsteroidal analgesics should not be used in pediatric animals. Opiates are the drug of choice, with their use restricted to the short-term and use balanced by reversal as needed.

Table 91-2. Side effects and signs of toxicity of commonly used drugs.

Drug	Signs of Toxicity
Amoxicillin/ampicillin	Rare: Gastrointestinal distress
Amoxicillin-clavulanate (Clavamox™)	Rare: Gastrointestinal distress
Cephalosporins	Rare: Gastrointestinal distress
Sulfadimethoxine (Albon™)	Rare: Crystalluria, keratoconjunctivitis sicca, bone marrow depression.
Pyrantel pamoate	Rare: Increased respiratory rate, ataxia
Metronidazole	Anorexia, vomiting, depression, mydriasis, neurologic signs
Praziquantel	Rare: Vomiting, ataxia, depression
Piperazine	Emesis, neurologic signs including muscle fasciculations, paralysis, death
Fenbendazole	Extremely rare
Thiopental	Central nervous system (CNS) and respiratory depression
Oxymorphone	CNS and respiratory depression, cardiovascular collapse
Propofol	Hypotension, bradycardia
Isoflurane/sevoflurane	Hypotension, respiratory depression, nausea, and vomiting

Post-procedure care and complications

Signs of toxicity must be evaluated carefully in neonatal animals and medications stopped if any such signs are noted (Table 91-2).

Supplemental reading

Little S. 2006. How I treat orphaned kittens. *Waltham Focus* 16:2–6.

Mathews KA. 2008. Pain management for the pregnant, lactating, and neonatal to pediatric cat and dog. *Vet Clin NA* 38:1291–1308.

Plumb DC. 2004. Drugs in neonates: Principles and guesses. Proceedings, Society for Theriogenology Annual Meeting, Lexington KY, pp. 307–315.

Plumb DC. 2008. *Veterinary Drug Handbook*. Ames, IA: Wiley-Blackwell.

Section X

Pediatric physical examination and management

92

What are normal physical examination findings at various ages in puppies and kittens?

General information

See Table 92-1 for a summary. Body weight should be measured at birth and daily thereafter. Normal birth weight in dogs varies from about 4 oz (120 g) in toy breeds to 22 oz (625 g) in giant breeds. Normal birth weight for kittens is about 3.5 oz (100 g). Body weight may decrease slightly within the first day of life but should remain stable or rise every day after that, doubling at 10 to 14 days of age.

Normal rectal temperature varies by age, averaging 96.0 ± 1.5 °F (35.6 ± 0.7 °C) in the first 7 days of life, 98.6 to 100.0 °F (37.0 to 38.2 °C) from 7 to 21 days of life, and gradually achieving adult values by 7 weeks of age.

Eyelids open at about 10 to 14 days on average. In some breeds, eyelids may open earlier. When the eyelids open, corneal edema may be present, evidenced by blue discoloration. This will clear within several days. Fundic examination can be performed as early as 6 weeks of age.

The ear canals are closed until 6 to 14 days of age. When they first open, the ear canals will contain cellular debris that will clear over the next several weeks. Otoscopic examination can be performed by 4 weeks of age. Hearing testing should be performed with the brainstem auditory evoked response (BAER) test; this requires specialized equipment and training. Use of percussion to assess hearing is not accurate because one may get either false negatives (animal inattentive) or false positives (animal responds to movement of hands or air movement).

Teeth erupt in a predictable manner, allowing use of dentition to determine age (Table 92-2). Oral mucous membranes may be quite hyperemic in normal puppies and kittens in the first week of life. Abdominal palpation can be used to assess size and abnormalities of the liver, left kidney, colon, and urinary bladder.

The skin and hair coat should be assessed for evidence of external parasites and dermatologic disease. The most common disorders of this type identified in puppies and kittens are fleas and dermatophytosis.

Auscultation of the heart and lungs is complicated by the small size of the animal relative to the size of the equipment available, and to the normal rapid heart rate of pediatric animals. Normal heart rate is 200 bpm in the first week of life. Bradycardia in neonates is not mediated by the vagal nerve and is commonly an indicator of hypoxia. Cardiac murmurs usually are ausculted at the base of the heart on the left side and vary in significance by grade. On a six-point scale, murmurs graded I to III most commonly are functional murmurs due to anemia,

Table 92-1. Timing of significant events in pediatric development.

Event	Age at Occurrence
Umbilical cord dries and falls off	2 to 3 days
Eyelids open	5 to 14 days
External ear canals open	6 to 14 days
Extensor dominance	5 days
Capable of crawling	7 to 14 days
Capable of walking, urinating, and defecating spontaneously	14 to 21 days
Hematocrit/red blood cell number stabilize near that of adult	8 weeks
Renal function nears that of adult	8 weeks
Hepatic function nears that of adult	5 months

Table 92-2. Timing of tooth eruption in dogs and cats (weeks).

Tooth Type	Dog		Cat	
	Deciduous	Adult	Deciduous	Adult
Incisor	2–3	12–16	3–4	12–16
Canine	3–4	16–20	3	16–24
Premolar	3–6	16–24	4–12	16–24
Molar	—	14–20	—	20–28

hypoproteinemia, fever, or septicemia. These innocent murmurs are more commonly identified in puppies than in kittens. Murmurs graded IV to VI most likely are due to congenital anomalies such as persistent ductus arteriosus. Normal respiratory rate is 10 to 35 breaths per minute.

Palpation of the skull may be performed to identify open fontanelles. There is no correlation in the literature between presence of open fontanelles, presence of hydrocephalus, and eventual signs of neurologic disease. The average date of closure of open fontanelles is not described.

Musculoskeletal examination can be performed at any time. Puppies and kittens should be able to hold their head up at birth, to push themselves up on their forelimbs and crawl by 2 weeks of age, and to walk by about 3 weeks of age.

Neurologic function does not mature until 6 to 8 weeks of age but neurologic examination can be performed at any age, localizing neurologic disease as in adults. At birth, puppies and kittens can suckle, vocalize, and respond to odor, touch, and pain. At birth, puppies and kittens have a functioning but slow withdrawal reflex and have flexor dominance. Extensor dominance becomes apparent within the first 3 weeks of life.

Clinical implications

Failure to gain weight may be the first sign of an abnormality of a puppy or kitten. Causes of failure to thrive are described in Chapter 96.

Hypothermia is associated with lack of intestinal motility, with ileus most common when body temperature falls to less than 94 °F (34.4 °C). Appetite also is suppressed in hypothermic animals, perhaps as a mechanism to minimize the risk of aspiration.

Abdominal distention may prevent ability to palpate abdominal organs. Distention most commonly is due to aerophagia, often secondary to persistent crying or increased respiratory effort; maldigestion; or retention of feces or urine. If the spleen is palpable, it is enlarged. If the liver margins are palpable beyond the rib cage, the liver is enlarged.

Functional cardiac murmurs (Grade I to III on a six-point scale) usually are due to anemia, hypoproteinemia, fever, or septicemia (see Chapter 98). Congenital murmurs (Grade IV to VI) are associated with clinical manifestations of dyspnea, cyanosis, open-mouth breathing, failure to suckle, and lethargy.

Radiography for assessment of bone and joint abnormalities is difficult in puppies and kittens because of decreased mineralization of bone. Quality of radiographs may be improved by using a tabletop cassette, and by decreasing kVp to one-half that for an adult of the same thickness or, in puppies, by using 2 kvP for each 1 cm of soft tissue measured for values up to 80 kVp.

Supplemental reading

Grundy SA. 2006. Clinically relevant physiology of the neonate. *Vet Clin NA* 36:443–459.

Lawler DF. 2008. Neonatal and pediatric care of the puppy and kitten. *Theriogenology* 70:384–392.

Little S. 2006. How I treat orphaned kittens. *Waltham Focus* 16:2-6.

Root Kustritz MV. 2006. *The Dog Breeder's Guide to Successful Breeding and Health Management*. St. Louis, MO: Elsevier, pp. 204–223.

93

What are normal values on complete blood count, chemistry profile, and urinalysis of puppies and kittens?

General information

Complete blood count

Changes seen on complete blood count (CBC; Table 93-1) include a decrease in hematocrit and an increase in polychromasia, nucleated RBCs, Howell–Jolly bodies, and Heinz bodies (kittens). Hematocrit reaches adult values by 8 weeks of age. A stress neutrophilia may be seen due to venipuncture. White blood cell number may be increased in normal pups up to 8 weeks of age.

Serum chemistry profile

See Table 93-2 for serum chemistry profiles in puppies and kittens. Liver enzyme values vary in pediatric animals, with a decrease in alanine aminotransferase concentrations and an increase in alkaline phosphatase (ALP) concentrations. ALP concentration remains elevated throughout the period of most rapid growth, which varies with adult size in dogs and peaks in kittens at 7 months of age, decreasing in both species as long bone physes close. Phosphorus mirrors this pattern. Total protein and albumin are decreased in young animals, reaching adult values by 6 to 9 months of age. Fasting and post-prandial bile acid concentrations are as adult concentrations by 4 weeks of age.

Renal function is immature in neonates, gradually achieving adult levels by 8 weeks of age. Glomerular filtration rate is reduced in puppies and kittens, varying from 20% of adult values at birth to 100% of adult values by several weeks of age. Blood urea nitrogen (BUN) values vary with time of sampling to ingestion of the most recent meal, but BUN is still a more sensitive indicator of renal function than is creatinine in young animals.

Normal neonates have relatively low blood glucose, averaging about 70 mg/dl from 3 days of age and gradually increasing to normal adult values. Significant hypoglycemia is defined as a blood glucose concentration of <50 mg/dl and often is associated with septicemia (see Chapter 97). Pathologic hyperglycemia is uncommon in pediatric patients; hyperglycemia may be evident in young animals as they near death.

Calcium and phosphorus concentrations are altered by slow update through the gastrointestinal tract and need for these minerals as a component of mineralizing bone. Phosphorus is elevated during rapid bone growth and is normal by 8 to 12 months of age; normalization may take longer in giant breeds.

Table 93-1. Complete blood count (CBC) values for puppies (P) and kittens (K).

Age	Hematocrit (%)		WBC (×10³/µl)		Differential (×10³/µl)											
	P	K	P	K	Neutrophils		Bands		Lymphocytes		Monocytes		Eosinophils		Basophils	
					P	K	P	K	P	K	P	K	P	K	P	K
1 day	36	—	—	—	—	—	—	—	—	—	—	—	—	—	—	—
7 days	37	—	—	—	—	—	—	—	—	—	—	—	—	—	—	—
2 weeks	29–53	34–37	7–23	9–10	3–10	5–7	0–1	0–0.1	2–7	3–4	0–1	0	0–2	0–2	0	0
4 weeks	27–37	26–27	9–26	14–17	4–13	6–8	0–0.3	0–0.2	1–8	6–7	0–2	0	0–1	1	0	0
6 weeks	26–36	26–28	13–27	16–19	4–18	8–11	0–0.3	0–0.3	3–17	6–7	0–3	0	0–1	1	0	0
8 weeks	31–39	29–31	13–17	16–20	6–12	6–8	0–0.3	0–0.3	3–7	8–11	0–2	0	0–1	1	0	0

WBC = white blood cell number.

Table 93-2. Serum chemistry profile values in puppies (P) and kittens (K).

Age	ALT (IU/l)		AP (IU/l)		Albumin (g/dl)		Total Protein (g/dl)		Glucose (mg/dl)		BUN (mg/dl)		Creatinine (mg/dl)		Sodium (mEq/l)		Chloride (mEq/l)		Potassium (mEq/l)		Calcium (mg/dl)	
	P	K	P	K	P	K	P	K	P	K	P	K	P	K	P	K	P	K	P	K	P	K
1 day	38	29–77	1069	1348–3715	2.3	1.9–2.7	4.0	3.9–5.8	106	65–149	33	34–94	0.6	0.6–1.2	147	—	104	—	5.8	—	11.5	9.6–12.2
7 days	22	11–76	242	126–363	2.3	2.0–2.5	4.3	3.5–4.8	129	105–145	24	16–36	0.4	0.3–0.7	148	—	111	—	6.1	—	11.8	10–13.7
2 weeks	10–34	10–24	176–560	116–306	2	2.1–2.6	4	3.7–5	111–146	107–158	—	11–30	—	0.4–0.6	—	—	—	—	—	—	—	9.9–13
4 weeks	20–22	14–55	135–201	90–274	1–2	2.4–2.9	4	4.5–5.6	86–115	99–152	—	10–22	—	0.4–0.7	—	149–153	—	120–124	—	4–5	—	10–12.2
6 weeks	16–17	—	125–132	—	4–5	2	3–4	4–5	125–126	<120	9	<30	1–4	0.6	148	151–156	105	119–125	5	5–6	11	10–11
8 weeks	9–24	12–56	144–177	60–161	2–3	2.4–3.0	4–5	4.8–6.5	134–272	94–143	—	16–33	—	0.6–1.2	—	150–152	—	119–125	—	4–5	—	9.8–11.7

ALT = alanine aminotransferase, AP = alkaline phosphatase, BUN = blood urea nitrogen.

Urinalysis

Because renal function is immature, urine specific gravity is decreased in pediatric animals, averaging 1.006–1.017. Physiologic proteinuria is present in the first days of life as colostral antibodies are absorbed and excreted in the urine. Normoglycemic glucosuria may be present until renal function matures.

Clinical implications

All bloodwork must be interpreted using pediatric values as normals. In general, by 12 to 20 weeks of age, all animals can be evaluated as adults.

Supplemental reading

Grundy SA. 2006. Clinically relevant physiology of the neonate. *Vet Clin NA* 36:443–459.

Harper EJ, Hackett RM, Wilkinson J, et al. 2003. Age-related variations in hematologic and plasma biochemical test results in Beagles and Labrador retrievers. *J Amer Vet Med Assoc* 223:1436–1442.

Lawler DF. 2008. Neonatal and pediatric care of the puppy and kitten. *Theriogenology* 70:384–392.

Levy JK, Crawford PC, Werner LL. 2006. Effect of age on reference intervals of serum biochemical values in kittens. *J Amer Vet Med Assoc* 228:1033–1037.

94

How are orphan puppies and kittens best fed?

General information

Orphan puppies and kittens must be provided with correct environmental temperature. Pediatric animals are incapable of thermoregulation until they are walking and can generate heat by muscle activity. Orphans should be maintained in an environment at about 85 °F (29.5 °C) for the first 2 weeks of life, and about 80 °F (26.5 °C) for the subsequent 2 weeks. Warmed and cooler areas should be available, so the puppies and kittens can crawl to an area appropriate in temperature. Among surface sources of heat, circulating hot water blankets are best. Hot water bottles should be wrapped in towels and frequently changed. Regular heating pads may heat unevenly and more easily burn the fragile skin of neonates. Heat lamps are an excellent, safe way to provide radiant heat; if neonates are not capable of crawling from warmer to cooler areas, they must be carefully observed.

It is important to determine, if possible, whether or not the pup or kitten ingested colostrum in the first 24 h of life. Maximal absorption of antibodies through the intestine occurs at 8 h after birth and decreases significantly by 1 day of life. If the owner is unsure whether a puppy has ingested colostrum, blood can be drawn from the puppy and from a littermate that is known to have nursed, and serum alkaline phosphatase and gamma-glutyl transpeptidase concentrations compared between the two. Concentrations remain high for only days in puppies that have ingested colostrum. If a puppy or kitten has not ingested colostrum, antibodies can be provided by subcutaneous administration of serum from the dam or another immunocompetent animal in the household. The empirical dose for kittens is 15 ml of serum, given subcutaneously as 5 ml boluses at birth and 12 and 24 h later. Donor cats should be negative for feline leukemia virus and feline immunodeficiency virus, and should have the same blood type as the kitten. The empirical dose for puppies is 10 ml/lb (22 ml/kg) of pooled adult serum; this can be given at once in large pups or split into boluses as described for kittens. Early vaccination may be warranted in puppies and kittens that did not receive adequate amounts of colostrum.

A milk-based diet should be offered for the first 3 to 4 weeks, with weaning begun at about 4 weeks of age. Urination and defection must be stimulated by genital manipulation after feeding until the animal is walking, at about 3 weeks of age.

Commercial milk replacers are recommended over homemade diets. Commercial formulas are balanced, of high nutrient density, low in fiber, and contain protein of high biologic value. Feline milk replacers must contain a source of taurine for optimal growth to occur. Homemade diets are best used in emergency situations only (Table 94-1).

Table 94-1. Emergency homemade milk replacers for puppies and kittens.

Puppies	Kittens
1. 3 parts evaporated milk (not skim) to 1 part water	1. 1/2 cup (120 ml) whole milk, 1 egg yolk, 1 drop liquid infant vitamins
2. 1 cup (240 ml) whole milk, 1 tsp (5 ml) vegetable oil, 1 drop liquid infant vitamins	2. 1/2 cup (120 ml) condensed milk, 1/2 cup (120 ml) water, 1/2 cup (120 ml) plain yogurt, 3 to 4 egg yolks
3. 1/2 cup (120 ml) whole milk, 1/2 cup (120 ml) water, 1 to 2 egg yolks, 2 Tums (calcium supplement), 1 tsp (5 ml) vegetable oil	
4. 1 cup (240 ml) whole milk, 1 T (15 ml) vegetable oil, pinch salt, 3 egg yolks, 1 drop liquid infant vitamins	

Amount to be fed varies with the caloric density of the formula, and age and weight of the animal. Kittens should receive 100 to 175 kcal/lb (220 to 380 kcal/kg) daily, split into four to six feedings. Puppies should receive 105 to 120 kcal/lb (230 to 260 kcal/kg) daily, split into four feedings. Less frequent feeding may be associated with volume overload, abdominal discomfort, diarrhea, and aspiration. Feeding frequency can be decreased to three times daily after the orphan reaches 2 weeks of age. For example, kitten homemade diet number 1 shown in the table contains 237 kcal/cup, which equals 30 kcal/fl oz (3 kcal/ml). Kittens require 100 to 175 kcal/lb daily; a 4-oz (1/4 lb = 113 g) kitten requires 25 to 44 kcal daily. The amount to be fed daily is 0.8 to 1.5 fl oz (8 to 15 ml). This yields, for four daily feedings, a per feeding volume of 0.2 to 0.4 oz (2 to 4 ml). Similarly, puppy homemade diet number 3 shown in the table contains 208 kcal/cup. Calculation determines a per feeding volume for a 1-lb puppy of 1 to 1.3 fl oz (10 to 12 ml). Commercial milk replacers often bypass discussion of calories, providing volume instructions only. Examples are the puppy milk replacer Esbilac and kitten milk replacer KMR (2 T/4 oz (10 ml/125 g) daily; see Resources).

The formula should be warmed to 95 to 100 °F (35 to 38 °C) when feeding 1- to 2-week-old animals. If the animal has been inappetent, it may be beneficial to feed half the calculated dose for 1 to 2 days. Equipment for feeding of pediatric animals includes spoons, droppers, bottles, or tubes. Using spoons and droppers to feed kittens and puppies is dangerous because the limited gag reflex of these animals easily permits aspiration of formula into the lungs. Bottle-feeding poses less risk of aspiration and more readily satisfies the neonate's need to suckle. Small bottles marketed for animals or bottles intended for premature human infants may be used. The hole in the nipple should allow milk to ooze slowly. The bottle should never be squeezed to force expulsion of milk while the animal is nursing.

Tube-feeding is quick. Caution must be taken to ensure proper placement of the tube into the gastrointestinal (GI) tract and to prevent overflowing and regurgitation. The animal should be held horizontally on its ventrum. The feeding tube varies in diameter and length with age. A #5 French feeding tube should be used in animals weighing less than 300 g, and a #8 to #10 French feeding tube should be used in animals weighing more than 300 g. Measure the length of the feeding tube by marking off 75% of the distance from the animal's last rib to the tip of its nose.

This length ensures placement in the stomach without kinking of the tube within the GI tract. Length should be rechecked and adjusted weekly. The warmed formula is gently expelled through the tube with a syringe. Monitor gastric distension; average stomach capacity in neonates is about 0.7 fl oz (4 tsp) per pound (40 ml/kg). Kink the tube before pulling it out to prevent leaking and aspiration of formula as the tube moves through the oral cavity.

Weaning, introduction of solid food, begins at 3 weeks of age in puppies and at 3 to 4 weeks in kittens. Puppy or kitten food, either canned food or moistened dry food, should be offered in preference to human foods. Food should be offered as a gruel initially, formed by thoroughly blending 1 part dry food to 3 parts water or 2 parts canned food to 1 part water for puppies, and 1 part dry food to 3 parts formula or 2 parts canned food to 1 part formula for kittens. Fresh water always should be provided as well. Gradually mix less water or formula with the food until the puppy or kitten is eating dry food exclusively. Weaning usually is complete by 6 to 8 weeks of age. By the time the animal is weaned, it should have a body weight roughly 6 to 10 times its birth weight.

Clinical implications

An effort should always be made to find another lactating bitch or queen to support orphan animals. No homemade or commercial milk replacer duplicates all the nutritional components and antibodies, enzymes, and other unique features of bitch or queen milk. Another important aspect of nursing from a foster mother is behavioral development; socialization by a dam from the same species is difficult to replace.

Problems reported with feeding of either commercial or homemade milk replacers to puppies and kittens include small, focal cataracts that develop because of deficiencies in vitamins or amino acids that resolve after weaning, and slower growth rate due to lack of enzymes necessary for fat digestion. By several months of age hand-raised puppies and kittens achieve the same size as littermates allowed to nurse. For commercial replacers based on cow's milk, the high lactose content and decreased caloric density make it difficult to provide orphans with an adequate number of calories without inducing diarrhea. If feeding induces diarrhea, the formula can be diluted 1 : 2 with electrolyte solution until the neonate can tolerate it.

Supplemental reading

Lawler DF. 2008. Neonatal and pediatric care of the puppy and kitten. *Theriogenology* 70(3):384–392.

Little S. 2006. How I treat orphaned kittens. *Waltham Focus* 16:2–6.

95

What flea and tick control products are safe to use in puppies and kittens?

General information

Commercially available products can be used on puppies and kittens as young as 4 to 6 weeks (Table 95-1).

Table 95-1. Flea and tick products suitable for use in puppies and kittens.

Name of Product	Drugs Contained	Route of Administration	Active Against …	Safe during Pregnancy?	Youngest Age Can Be Used	
					Puppies	Kittens
Frontline	Fipronil	Topical	Adult fleas, ticks	Not approved	10 weeks	8 weeks
Sentinel	Lufenuron + milbemycin	Oral	Flea eggs and larvae, heartworm, rounds, hooks	Safe	4 weeks (or 2 lb)	Not for use in cats
Advantage	Imidacloprid	Topical	Adult fleas	Not approved in the United States but used elsewhere	7 weeks	8 weeks
Advantix	Imidacloprid + permethrin	Topical	Adult fleas, ticks, mosquitoes	Not approved	7 weeks	Not for use in cats
Revolution	Selamectin	Topical	Flea eggs and larvae, adult fleas, ticks, ear mites, mange, hooks, heartworm	Safe	6 weeks	6 weeks
Capstar	Nitempyram	Oral	Adult fleas	Manufacturer claims safe	4 weeks (or 2 lb)	4 weeks (or 2 lb)
Program	Lufenuron	Oral	Flea eggs and larvae	Safe	6 weeks	6 weeks

PEDIATRIC EXAMINATION

Clinical implications

Flea control is desirable but difficult in very small animals. The feeding of fleas, coupled with normal physiologic changes in hematocrit in young animals, can precipitate pathologic anemia.

Dips are not recommended because of the difficulty in gauging the amount of drug taken up on application and through later exposure. Shampoos may be less toxic because they have no residual activity. Flea combs can be used to some effect in very young animals.

Section XI

Pediatric disease

96

What are the causes of failure to thrive in puppies and kittens?

Etiology

Failure to thrive, or fading, is a poorly defined phenomenon. It is defined by some as failure of weight gain and general unthriftiness from birth, and by others as regression in growth and development in animals that had been growing well. There is no one etiology associated with either phenomenon.

Differentials common to both groups include low birth weight; abnormal environment or maternal neglect with secondary hypothermia; failure to ingest colostrum within 24 h of birth; parasites (see Chapter 97); and septicemia (see Chapter 98). Differentials for animals that never grow well include failure to suckle effectively (cleft palate or abnormal nipples), congenital abnormalities of metabolism, and, in cats, neonatal isoerythrolysis (see Chapter 99). Differentials for animals that had been growing well but then stagnate or regress in growth and development vary depending on body condition. Differentials for those with poor body condition include portosystemic shunt, renal failure, megaesophagus, exocrine pancreatic insufficiency, and cardiac disease. Differentials for those with good body condition include hypothyroidism, diabetes mellitus, and adrenal disease; these are very uncommon in puppies and kittens.

In general, one can identify environmental, genetic, and infectious causes of failure to thrive (Table 96-1). Of the many infection-related possibilities, only canine herpesvirus will be covered in detail in this section; puppies that acutely stop nursing, cry incessantly, and die within 24 h of first clinical signs may be infected with canine herpesvirus.

Clinical signs

Puppies and kittens should be seen if they are not gaining weight. The owner should weigh them at birth and every day thereafter; puppies and kittens should remain stable or gain weight every day, doubling their birth weight by about 10 days of age. Other nonspecific signs of illness include moving away from the dam and littermates or being persistently rejected by the dam, crying for more than 20 min, and poor muscle tone.

Cleft palate may be evidenced by passage of milk through the nostrils while suckling or by respiratory distress if milk is aspirated into the lungs. Clinical manifestations of septicemia are variable (see Chapter 98). Neonatal isoerythrolysis is evidenced by lethargy and icterus (see Chapter 99). Clinical signs associated with other disorders are as in adults.

Table 96-1. Causes of failure to thrive in puppies and kittens.

General Cause	Specific Cause[a]
Environment	Hypothermia
	Hyperthermia
	Maternal factors
	Environmental toxins
Genetic	Gross developmental abnormalities
	Thymic dysfunction
Infectious agents—Bacteria	*Brucella canis* (P)
	Bartonella henselae (K)
	Campylobacter sp. (P/K)[b]
	Gram-negative enterobacters (P/K)[c]
	Gram-positive cocci (P/K)[c]
Infectious agents—Viruses	Canine herpesvirus (P)
	Canine adenovirus (P)
	Canine parvovirus types 1 and 2 (P)[b]
	Canine distemper virus (P)
	Feline herpesvirus type 1 (rhinotracheitis) (K)
	Calicivirus (K)
	Feline leukemia virus (K)
	Coronavirus (K)
	Feline immunodeficiency virus (K)
	Panleukopenia virus (feline distemper) (K)
Infectious agents—Parasites	Roundworms (*Toxocara* sp.) (P/K)[b]
	Hookworms (*Ancylostoma caninum*) (P)[b]
	Toxoplasma gondii (K)
	Tritrichomonas foetus (K)[b]
	Giardia sp. (P/K)[b]
	Coccidia (P/K)[b]

[a]P, puppies; K, kittens.
[b]See Chapter 97.
[c]See Chapter 98.

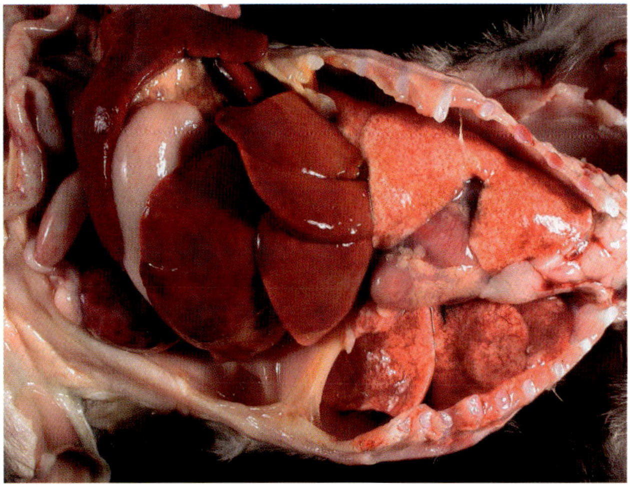

Figure 96-1: Typical lesions of herpesvirus on abdominal organs (reprinted with permission from Root Kustritz MV. 2006. *The Dog Breeder's Guide to Successful Breeding and Health Management,* St. Louis, MO: Elsevier).

Canine herpesvirus may be apparent late in gestation as pregnancy loss (see Chapter 46) or as birth of a mixed litter of stillborn, macerated, mummified, and apparently normal puppies. Pups exposed within the first 3 weeks of life develop acute viremia with subsequent vasculitis of all internal organs (Fig. 96-1). The most common description is acute onset of lethargy, crying, and unwillingness to suckle in pups that had been vigorous and growing rapidly. Affected pups often die within 24 to 48 h of onset of clinical signs.

Diagnosis

Normal rectal temperature varies by age, averaging 96.0 ± 1.5 °F (35.6 + 0.7 °C) in the first 7 days of life, 98.6 to 100.0 °F (37.0 to 38.2 °C) from 7 to 21 days of life, and gradually achieving adult values by 7 weeks of age.

If the owner is unsure whether a puppy or kitten has ingested colostrum, blood can be drawn from the neonate of interest and a littermate that is known to have nursed, and serum alkaline phosphatase and gamma-glutyl transpeptidase concentrations compared between the two. Concentrations remain high for only days in puppies or kittens that have ingested colostrum.

Cleft palate is diagnosed by inspection (Fig. 96-2). Diagnosis of septicemia is discussed in Chapter 98. Diagnosis of neonatal isoerythrolysis is discussed in Chapter 99. Diagnosis of other disorders is as in adults.

Canine herpesvirus infection in pups most often is made at necropsy. Diffuse pinpoint hemorrhages of the kidney, liver, and intestinal tract are pathognomonic. Free fluid may be present in the abdominal and thoracic cavities. Inclusion bodies may be identified in hepatocytes. Virus isolation and polymerase chain reaction (PCR) testing also is available.

Treatment

Rewarming of hypothermic neonates should be gradual, taking anywhere from 30 min to 2 h. Too rapid warming may be associated with increased metabolic demand. Remember that normal body temperature for animals less than 3 weeks of age is lower than that for adults. The use of

Figure 96-2: Cleft palate.

indirect heat, such as a radiant heat lamp, is preferred to the use of direct heat, such as a heating pad. Body temperature less than 94 °F (34.4 °C) is associated with ileus, or gastrointestinal stasis. Hypothermic animals should not be fed milk-based products, which will ferment in the stomach if motility is absent.

If a puppy or kitten has not ingested colostrum, antibodies can be provided by subcutaneous administration of serum from the dam or another immunocompetent animal in the household. The empirical dose for kittens is 150 ml/kg or 15 ml of serum, given subcutaneously as 5 ml boluses at birth and 12 and 24 h later. Donor cats should be negative for feline leukemia and feline immunodeficiency virus, and should have the same blood type as the kitten. The empirical dose for puppies is administration of 10 ml/lb (22 ml/kg) of pooled adult serum; this can be given at once in large pups or split into boluses as described for kittens.

Puppies and kittens dehydrate quickly because of their relatively large surface to volume ratio compared to adults, greater skin permeability, increased total body water content, and decreased renal function with subsequent inability to retrieve water. Dehydrated neonates that are not hypothermic can be tubed and provided with oral fluids. Ill neonates require parenteral therapy. Fluids can be administered into the subcutaneous space but are taken up variably. Fluids containing dextrose cannot be given subcutaneously as this hypertonic solution actually may draw more fluid into that space. Intravenous catheters (20- to 22-gauge cephalic catheters) can be placed; in moribund animals, intraosseous catheter placement may be considered (see Chapter 90).

Cleft palates may be closed surgically or permitted to close spontaneously. Treatment of septicemia is discussed in Chapter 98. Treatment of neonatal isoerythrolysis is discussed in Chapter 99. Treatment of other disorders is as in adults.

Canine herpesvirus is difficult to treat. The virus does not replicate well at normal to high body temperatures, so some advocate maintaining the pups in a high environmental temperature such that their body temperature is 101.0 to 102.2 °F (36.3 to 37.0 °C). The antiviral agent, acyclovir, has been described for therapy but has not been evaluated in scientific studies. One described dose is 10 mg/kg every 6 h per os for 5 days.

Supplemental reading

Addie DD. 2000. Prevention of infectious causes of death in pups and kittens. Proceedings, British Small Animal Veterinary Association, Bromsgrove, Worcestershire, England, pp. 1–6.

Crawford PC, Levy JK, Werner LL. 2006. Evaluation of surrogate markers for passive transfer of immunity in kittens. *J Amer Vet Med Assoc* 228:1038–1041.

Freshman JL. 2005. Causes of fading puppy and kitten syndrome. *Vet Med* 100:781–788.

Freshman JL. 2005. Initially treating fading puppies and kittens. *Vet Med* 100:800–805.

Lawler DF. 2008. Neonatal and pediatric care of the puppy and kitten. *Theriogenology* 70:384–392.

Levy JK, Crawford PC, Collante WR, Papich MG. 2001. Use of adult cat serum to correct failure of passive transfer in kittens. *J Amer Vet Med Assoc* 219:1401–1405.

Macintire DK. 2008. Pediatric fluid therapy. *Vet Clin NA* 38:621–627.

Root Kustritz MV. 2003. *The Practical Veterinarian: Small Animal Theriogenology*. St. Louis, MO: Butterworth-Heinemann, pp. 283–329.

97

What are the common causes of diarrhea in puppies and kittens?

Etiology

The most common cause of diarrhea in puppies and kittens is internal parasites. Common parasites are protozoa (Coccidia and Giardia) and helminths (roundworms and hookworms). Parasites are present in greater numbers in very young animals than in older animals; in one study, 10.9% and 7.8% of dogs less than 6 months of age had a positive fecal flotation test for roundworms and hookworms, respectively. Other causes of diarrhea include overnutrition or feeding of high-solids formulas in nursing pups and kittens, and supplementation with inappropriate foodstuffs in weaned puppies and kittens. Any cause of diarrhea in adult dogs and cats may also be a cause of diarrhea in puppies and kittens. Viral causes, including parvovirus, will not be discussed here in detail.

Clinical signs

In the first 3 weeks of life, the dam stimulates defecation and may mask presence of diarrhea. Dehydration and weight loss may be the first evidence of a problem. Animals with a heavy parasite burden also may have a pot-bellied appearance and scruffy hair coat. Roundworm larvae may migrate through the body, causing a nonproductive cough and poor weight gain. Large bowel diarrhea in kittens may be evident in animals carrying *Tritrichomonas foetus*.

Diagnosis

Fecal flotation may be used to identify the oocytes of roundworms and hookworms. Coccidia and *Giardia* sp. may be best identified on fresh fecal samples. The small size of puppies and kittens and grooming behavior of the queen may preclude ability to definitively diagnose presence of specific parasites. Fecal flotation on a sample from the dam may be informative. Fecal culture may be necessary to identify persistent infection with *Salmonella* sp., *Campylobacter* sp., or other bacterial causes of diarrhea. As with parasitism, fecal culture from the dam may inform decisions regarding possible bacterial infection as a cause of diarrhea in her offspring. Identification of *Tritrichomas foetus* requires special culture media or polymerase chain reaction testing; call your diagnostic laboratory for details of testing.

Treatment

For roundworms and hookworms, puppies or kittens aged 2 weeks or older can be treated with pyrantel pamoate (5 to 10 mg/kg per os and repeat in 2 to 3 weeks). Coccidia and *Giardia* sp.

usually cause mild, self-limiting diarrhea. If treatment for coccidiosis is necessary, puppies and kittens can be treated with sulfadimethoxine (30 mg/kg once daily or 15 mg/kg twice daily in puppies, 30 mg/kg once daily in kittens weighing at least 1 kg, until signs regress). If treatment for giardiasis is necessary, puppies and kittens can be treated with fenbendazole (50 mg/kg once daily per os for 3–7 days) and perhaps with metronidazole, depending on age (see Chapter 91). *Tritrichomas foetus* infection in kittens can be treated with ronidazole (30 to 50 mg/kg twice daily for 2 weeks).

Parasite burden may be reduced in bitches and shedding at the time of whelping reduced by treatment with fenbendazole late in gestation (50 mg/kg once daily per os from about day 40 of gestation through early lactation).

Supplemental reading

Freshman JL. 2005. Initially treating fading puppies and kittens. *Vet Med* 100:800–805.

Magne ML. 2006. Selected topics in pediatric gastroenterology. *Vet Clin NA* 36:533–548.

Mohamed AS, Moore GE, Glickman LT. 2009. Prevalence of intestinal nematode parasitism among pet dogs in the United States (2003–2006). *J Amer Vet Med Assoc* 234:631–637.

Munnich A, Lubke-Becker A. 2004. Escherichia coli infections in newborn puppies—Clinical and epidemiological investigations. *Theriogenology* 62:562–575.

Root Kustritz MV. 2003. *The Practical Veterinarian: Small Animal Theriogenology*. St. Louis, MO: Butterworth-Heinemann, pp. 283–329.

98

What are the clinical manifestations of septicemia and how is it diagnosed and treated?

Etiology

Septicemia is, by definition, whole body infection with circulating bacteria. Animals that are hypothermic or that failed to ingest colostrum are predisposed to septicemia. Gram-negative organisms are most commonly isolated, with *Escherichia coli* reported as the most common causative organism. Other reported organisms include *Staphylococcus* sp., *Streptococcus* sp., *Klebsiella* sp., *Pseudomonas* sp., *Pasteurella* sp., *Enterobacter* sp., *Enterococcus* sp., *Clostridium* sp., *Bacteroides* sp., *Fusobacterium* sp., *Brucella canis*, and *Salmonella* sp.

The subtype of bacterium often is the same as that isolated from the vulvar discharge of the bitch. The most common route of entry is the umbilicus, but neonates also may be infected by exposure through the gastrointestinal tract, respiratory tract, urinary tract, and skin lacerations.

Clinical signs

Clinical signs vary with organ(s) infected. Possibilities include vomiting foamy fluid, passing liquid diarrhea, and reddening of the anus; pyelonephritis with abdominal pain, hematuria, and fever; respiratory distress, open-mouth breathing, and cyanosis; conjunctivitis; and omphalitis (inflammation of the umbilicus). Nonspecific signs include lethargy, lack of suckling, dehydration, and sloughing of the extremities secondary to vasculitis and tissue hypoxemia (Fig. 98-1).

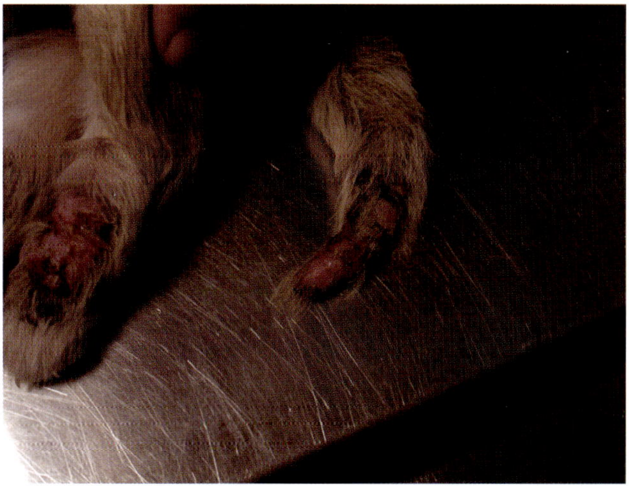

Figure 98-1: Ischemic skin lesions of the extremities secondary to vasculitis in a puppy with septicemia.

Diagnosis

Definitive diagnosis requires blood culture. However, the relatively large volume of blood required is prohibitive in very small animals. One reported technique is dilution of 1 ml of whole blood into 5 to 10 ml enrichment broth, with any growth by 6 to 18 h later considered diagnostic. Culture of urine may be positive (see Chapters 88 and 89 for venipuncture and urine collection technique). Supportive diagnostics include normocytic, normochromic anemia, and mild to moderate neutrophilia on complete blood count, and hypoglycemia on serum chemistry profile.

Septicemia often is diagnosed at necropsy. There are no pathognomic signs of septicemia, but pathology indicative of infection will be evident in infected organ systems. Culture of tissue or free abdominal or thoracic fluid is definitive.

Treatment

Antibiotic choice should be based on culture and sensitivity, if possible, and specifics of pediatric physiology must be considered (see Chapter 91). Penicillins and cephalosporins are logical empirical choices.

Supportive therapy includes fluid replacement with a balanced electrolyte solution. Dextrose (5%) may be added, and KCl supplemented if serum potassium concentrations are less than 2.5 mEq/L. Oxygen therapy may be necessary for management of tissue hypoxemia. Provision of prebiotics that support growth of the intestinal microflora may be beneficial.

Supplemental reading

Munnich A, Lubke-Becker A. 2004. Escherichia coli infections in newborn puppies—Clinical and epidemiological investigations. *Theriogenology* 62:562–575.

Root Kustritz MV. 2003. *The Practical Veterinarian: Small Animal Theriogenology*. St. Louis, MO: Butterworth-Heinemann, pp. 283–329.

99

How is neonatal isoerythrolysis diagnosed, treated, and prevented in cats?

Etiology

Cats have two main blood types, A and B. Cats with type A red blood cell (RBC) antigens have weak anti-B antibodies; cats with type B blood have very strong anti-A antibodies. These antibodies develop naturally, not requiring prior transfusion or pregnancy as exposure. If a type B queen is bred by a type A tom, she may produce kittens who are type A. When those kittens ingest colostrum, the strong anti-A antibodies ingested will lyse their RBCs, precipitating a hemolytic crisis.

Clinical signs

Affected kittens are icteric and anemic and may slough the tip of the tail and extremities as blood flow to those areas is compromised. Kittens demonstrate tachypnea and tachycardia, and hemoglobinuria, and may die acutely. Clinical signs usually appear within the first days of life.

Diagnosis

Neonatal isoerythrolysis usually is diagnosed on the basis of clinical signs and age at which they appear.

Treatment

Kittens should be removed from the queen if they are less than 1 day of age and still capable of taking up antibodies across the gastrointestinal (GI) tract as they ingest the queen's colostrum. They can be placed back on the queen at 2 to 3 days of age as they are no longer capable of absorbing those large proteins across the GI tract at that point in development. Transfusion may be required. The goal is to provide the kitten with RBCs to permit delivery of oxygen to tissues until they can build up a store of their own RBCs. Because the dam's antibodies do not attack her own RBCs, she is a good source of RBCs for transfusion to the kitten. Other alternatives include transfusion of washed type B blood cells or transfusion from a donor that has been cross-matched with the kitten. All donors should be negative for feline leukemia virus and feline immunodeficiency virus. Recommended administration regimen for blood products is 10 to 20 ml/kg over a 4-h period, via an intravenous or intraosseous catheter. Mortality rate is high.

Prevention is preferred to treatment. All cats should be blood-typed before being used for breeding and like bred to like. There also is variation by breed; virtually all domestic short-haired cats in the United States are type A. The type B blood type is most common in the Devon Rex, Cornish Rex, and British Shorthair breeds. Likelihood of incompatibility of mating is estimated at 14 to 25% for Persians and Abyssinians. Umbilical blood from kittens also may be used for blood typing, before those kittens are allowed to nurse from the dam.

Supplemental reading

Addie DD. 2000. Neonatal isoerthyrolysis. Proceedings, British Small Animal Veterinary Association, Bromsgrove, Worcestershire, England.

Freshman JL. 2005. Initially treating fading puppies and kittens. *Vet Med* 100:800–805.

Root Kustritz MV. 2003. *The Practical Veterinarian: Small Animal Theriogenology*. St. Louis, MO: Butterworth-Heinemann, pp. 283–329.

100

What are the causes of stillbirths and neonatal mortality in kittens and puppies?

Etiology

Mortality rate in the first weeks of life is high in puppies and kittens. The rate of stillbirths in cats ranges from 4.7 to 11.2%. In one study of 421 pups that died by day 45 of life, 29.9% were stillborn. Mortality rate is highest in the first 7 days of life, averaging 27.3% in kittens and 26.0% in puppies.

Causes of stillbirth and neonatal mortality in puppies include intrauterine bacterial infection, canine herpesvirus (see Chapter 96), genetic abnormalities, savaging by the dam, and maternal neglect. Increased puppy mortality is associated with increasing age of the dam. Causes of stillbirth and neonatal mortality in kittens include intrauterine bacterial infections, genetic abnormalities, neonatal isoerythrolysis (see Chapter 99), and savaging by the dam. Increased kitten mortality is associated with increased parity of the queen, obesity of the queen, and litter size of one or seven or more kittens. Predisposed breeds include Persians, Manx, and Himalayans.

Clinical signs

Clinical manifestation in the dam may help identify causes of stillbirth or neonatal mortality, as may history regarding number in the litter affected. Intrauterine bacterial infection often is associated with purulent vulvar discharge from the dam and loss of multiple offspring. Canine herpesvirus often is associated with birth of a mixed population of apparently normal and stillborn pups (see Chapter 96). Maternal neglect may be due to pain, as in dams with mastitis. Puppies and kittens that are persistently pushed away by the dam should be removed and hand-raised. Savaging is anecdotally associated with hypocalcemia (see Chapter 40) and may be more evident in primiparous dams and in high-strung dams.

Diagnosis

The dam always should be evaluated and a complete history and physical examination completed. If abnormal vulvar discharge is present, a sample should be collected for culture and sensitivity and the dam tested for brucellosis if that was not done pre-breeding. Serum calcium concentrations should be evaluated.

Necropsy of dead puppies and kittens is strongly recommended. The dead neonate should be placed in a zip-top plastic bag and refrigerated, not frozen, before submission. Although a definitive diagnosis may be arrived at in only about one-third of cases, that diagnosis often changes management in the facility significantly. The expense of the necropsy usually is less than the continuing expense in puppies or kittens lost.

Treatment

Treatment is directed at the dam and littermates and is dependent on cause, if identified. Bitches and queens exhibiting poor mothering behavior most likely are not secreting significant amounts of oxytocin as their offspring fail to stimulate release by manipulation of the mammary gland. Oxytocin has been demonstrated to increase pair-bonding in some species, and administration of oxytocin may be useful in bitches and queens.

Supplemental reading

Elmaz O, Aksoy OA, Zonturlu A, et al. 2008. The determination of growth performance and some morphological characteristics effective on development curves of German Shepherd puppies during the suckling period. *Pol J Vet Sci* 11:367–370.

Root Kustritz MV. 2003. *The Practical Veterinarian: Small Animal Theriogenology*. St. Louis, MO: Butterworth-Heinemann, pp. 283–329.

Sparkes AH, Rogers K, Henley WE, et al. 2006. A questionnaire-based study of gestation, parturition and neonatal mortality in pedigree breeding cats in the UK. *J Fel Med Surg* 8:145–157.

Resources

The following is a list of companies and products that can be used in small-animal theriogenology. This list is not inclusive of all such products available, and inclusion on this list is not an endorsement of quality.

Equipment	Company
Cystoscopy (endoscopy) equipment for transcervical insemination of bitches	Minitube of America, Verona, WI, http://www.minitube.com
Cystoscopy (endoscopy) equipment for transcervical insemination of bitches	Karl Storz Endoscopy, Culver City, CA, http://www.karlstorz.de/cps/rde/xchg/SID-AA16E478-F17E7354/karlstorz-en/hs.xsl/2274.htm
Chilled semen breeding kit for dogs—Fresh Express	Synbiotics, Kansas City, MO, http://www.synbiotics.org
Chilled semen breeding kit for dogs—Puppy Pak	International Canine Semen Bank, Sandy, OR, http://www.ik9sb.com/Puppy_Pak.asp
Chilled semen breeding kit for dogs—Fresh Cooled Canine Semen	Camelot Farms, College Station, TX, http://www.camelotfarms.com/shipping_semen.php
Chilled semen breeding kit for dogs—Chilled semen kit	CLONE, Chester Springs, PA, http://www.cloneusa.com/chilled.html
Semen shipment container	Equitainer; Hamilton Research, South Hamilton MA, http://www.equitainer.com
Kit for shipment of chilled semen to be frozen—Cryo-Kit	International Canine Semen Bank, http://www.ik9sb.com/Cryo-Kit.asp
Rapid slide agglutination test (RSAT) for canine brucellosis	D-Tec™; Synbiotics, Kansas City, MO, http://www.synbiotics.com
Agar gel immunodiffusion (AGID) test for canine brucellosis	New York State Animal Health Diagnostic Center, Cornell University, Ithaca, NY, 607-253-4136

Equipment	Company
Agar gel immunodiffusion (AGID) test for canine brucellosis	University of Georgia College of Veterinary Medicine, Athens, GA, 229-386-3340
Karyotyping (chromosome analysis)	Texas A&M University, Molecular Cytogenetics Laboratory, College Station, TX, 979-458-0519
Hypo-osmotic swelling test for semen evaluation	Sigma Chemical, St. Louis, MO, http://www.sigmaaldrich.com
Formula for feeding of orphan kittens	KMR; Pet-Ag, Elgin, IL
Registration of eye certification by a board-certified veterinary ophthalmologist	Canine Eye Registry Foundation (CERF), http://www.vmdb.org/cerf.html
Registration of hip, elbow, heart, and thyroid status	Orthopedic Foundation for Animals (OFA), Columbia, MO, http://www.offa.org
Registration of hip joint laxity	Penn-HIP, Philadelphia, PA, http://www.pennhip.org
Registration of commonly used sires or sires with semen chilled or frozen	American Kennel Club (AKC), http://www.akc.org/dna/fus_faq.cfm
Computerized semen evaluation system	IVOS Sperm Analyzer; Hamilton-Thorne, Beverly, MA, http://www.hamiltonthorne.com/products/casa/ivos.htm
Computerized semen evaluation system	SpermaCue; Minitube of America, Verona, WI, http://www.minitube.com
Insemination pipettes for bitches	Reproduction Resources, Walworth, WI, http://www.reproductionresources.com
Monitoring during parturition	Whelp Wise; Veterinary Perinatal Specialties, Wheat Ridge, CO, 1-888-281-4867 or http://www.whelpwise.com
Intratesticular injection to induce sterility in dogs	Esterisol; Ark Sciencies, Baltimore, MD, http://www.arksciences.com/products/html
Rubber collecting cone for canine semen collection	Nasco, Fort Atkinson, WI, http://www.enasco.com/farmandranch/
Polypropylene collecting cone for canine semen collection	Lane Manufacturing, Denver, CO, http://www.lane-mfg.com/bovineprod.html

Index

Page numbers in *italics* refer to Figures, those in **bold** to Tables.

abdominal distention in pediatric animals, 279
abdominal palpation, use in pregnancy diagnosis
in the bitch, **22**
in the queen, **192**
abdominal radiographs, use in pregnancy diagnosis
in the bitch, **22,** *24*
in the queen, **192**
acute-phase proteins, use in pregnancy diagnosis in the bitch, **23**
agalactia in the bitch, 155
agarose gel immunodiffusion (AGID) test for canine brucellosis, **140,** 140–141
age and small litter size in the bitch, **108**
agglutination test for canine brucellosis, 139–140, **140,** *140*
AGID test. *See* agarose gel immunodiffusion (AGID) test
aglepristone, use in pregnancy termination
in the bitch, **113**
in the queen, **234**
AI. *See* artificial insemination
AKC. *See* American Kennel Club
Alizine™. *See* aglepristone
alkaline phosphatase (ALP) in canine seminal fluid, **76,** 76–77, 182
ALP. *See* alkaline phosphatase
American Kennel Club (AKC)
certification of freezing centers, 47
DNA sample requirement, 43, 47
anatomic tests for hereditary conditions, canine, 91, 92
androgens, use in contraception
in the bitch, 88
in the queen, 225–226
anesthesia technique for Cesarean section
in the bitch, 59–62, **60, 61**
in the queen, 203–205, **204**
anestrus
persistent
in the bitch, 165–167, **166**
in the queen, 253–254
stage in the estrous cycle
in the bitch, 7, *10,* 71, 73
in the queen, 213
anovulatory cycle, 170, 259
antibiotics, use in the bitch during heat, 95–96
artificial insemination (AI)
in the bitch
and small litter size, **108**
surgical intrauterine, 47, 57–58, *58*
timing, 100
transcervical, 43, 47, 55–56
vaginal, 43, 47, 51–54, *52, 53*
in the queen
surgical intrauterine, 201
vaginal, 199
aspermia, canine, 182
asthenozoospermia, canine, 183
automated semen evaluation systems, 32, 196
azoospermia, canine, 182

balanoposthitis, canine, 179, 180
behavioral changes in the bitch due to heat, 99

benign prostatic hypertrophy and hyperplasia (BPH), canine, 173–175, *174*
biopsy technique, canine
prostatic, 37–38
testicular, 41–42, *42*
blood types, feline, 227, 303–304
body weight in pediatric animals, 277, 293
bone abnormalities in pediatric animals, assessment of, 279
BPH. *See* benign prostatic hypertrophy and hyperplasia
breeding management
in the bitch
and behavioral/physical changes, 99
and electrical impedance, 99
and ferning of vaginal fluid, 99
and glucose concentration in vaginal fluid, 99
and luteinizing hormone assays, 97
and progesterone assays, 97, **98,** 99–100
and ultrasound evaluation of ovaries, 99
and vaginal cytology, 7, *8–10,* 97–98, 99
and vaginoscopy, 98
in the queen
use of progesterones and other measures, 229–230
and vaginal cytology, 189–190, *189, 190*
bromocriptine, use in pregnancy termination in the bitch, **112**

cabergoline
use in inhibiting lactation in the bitch, 73, 83
use in pregnancy termination
in the bitch, **112**
in the queen, **234**
calcium
gluconate, for treatment of hypocalcemia in the bitch, 130
loss of, and hypocalcemia in the bitch, 129
supplements in the bitch, 105, 106, 107, 129
calico male cat, 261
cancers, high-morbidity
and canine castration, 86
and canine ovariohysterectomy, 82
canine brucellosis, 139–142
causes, 139
clinical signs, 139
diagnosis, 139–141, **140,** *140*
management of *Brucella canis* positive animals, **141**
treatment, 141–142, **141**
Canine Eye Registry Foundation (CERF), and tests for hereditary conditions, 43, 47, 51, 55, 57, 92
canine herpesvirus (CHV)
and failure to thrive in puppies, 293, 295, *295,* 296
and pregnancy loss, 146, 147
cardiac murmurs in pediatric animals, 277–278, 279
CASA systems. *See* computer-assisted semen analysis systems
castration
canine
best age for, 85–86
and cranial cruciate ligament injury, 85–86
and high-morbidity cancers, 86
and obesity, 85
and prostate disease, 85, 174
and sexually dimorphic behavior, 85
and testicular neoplasia, 85
feline, 226
best age for, 223
and obesity, 223
and persistent male behavior, 257–258, *257*
catheter, intraosseous, technique for placement of in pediatric animals, 269–270, *269*
CBC. *See* complete blood count
CCL injury. *See* cranial cruciate ligament (CCL) injury
CEH. *See* cystic endometrial hyperplasia
CERF. *See* Canine Eye Registry Foundation
Cesarean section
in the bitch
anesthesia for, 59–62, **60, 61**
concurrent with ovariohysterectomy, 63, 64, 65, 83
and dystocia, **127,** 128
technique for, 63–65, *64*
in the queen
anesthesia for, 203–205, **204**
concurrent with ovariohysterectomy, 207, 208
and dystocia, 241
technique for, 207–208, *208*
Cheque™, 88, 225–226
chilled semen, preparation of, 43–45
CHV. *See* canine herpesvirus
cleft palate, 293, 295, *296,* 296
cloprostenol, use in pregnancy termination in the bitch, **112**
color, in semen evaluation
canine, 29, *30,* 75, **76,** *76*
feline, 195, 215, **215**
colostrum, in pediatric animals, 285, 295, 296
complete blood count (CBC), normal values, in pediatric animals, 281, **282,** 284

computer-assisted semen analysis (CASA) systems, 32, 196
concentration, in semen evaluation
canine, 29–31, *31,* 75–76
feline, 195–196, **215,** 216
contraception, nonsurgical alternatives
canine, 87–90
feline, 225–226
corticosteriods, use in pregnancy termination in the bitch, **113**
cranial cruciate ligament (CCL) injury
and canine castration, 85–86
and canine ovariohysterectomy, 82
cryopreservation, semen, 48
cryptorchidism, canine
as heritable defect, 119–120, 121, 122
locating without surgery, 119–120
medical therapy for, 121–122
C-section. *See* Cesarean section
cystic endometrial hyperplasia (CEH)
in the bitch, 147, 149
in the queen, 244, 245
cystoscopes, use of, 13, *14*
cytology
use in canine semen evaluation, 32
vaginal. *See* vaginal cytology
of vulvar discharge, in diagnosis of pyometra in the bitch, *151*

developmental abnormalities and pregnancy loss
in the bitch, 147
in the queen, 244
dexamethasone, use in pregnancy termination in the bitch, **113**
diarrhea in pediatric animals, 299–300
diestrus stage in the estrous cycle
in the bitch, 7, *10,* 71, 73
in the queen, 213
dietary recommendations
for orphan pediatric animals, 285–287, **286**
for pregnancy in the bitch, 105, 106, 107, 129
dietary supplements
for improvement of canine semen quality, 117–118
for improvement of fertility in the bitch, 105–106
DNA sample requirement, American Kennel Club, 43, 47
DNA-based tests for hereditary conditions, canine, 91, 92
Dostinex™. *See* cabergoline
drugs
pharmacokinetics of commonly used drugs for pediatric animals, **272,** 273
safety for use in pregnancy
in the bitch, 101, **102–103**
in the queen, 231, **231–232**
side effects of commonly used drugs for pediatric animals, **273**
signs of toxicity of commonly used drugs for pediatric animals, **273,** 274
use in anesthesia of the bitch during Cesarean section, **60**
use in pediatric animals, 271–274, **272, 273**
use in termination of pregnancy
in the bitch, 111, **112–113**
in the queen, 233, **234**
dystocia
in the bitch, 125–128
causes, 107, 125–126
and Cesarean section, **127,** 128
clinical signs, 126
diagnosis, 126–127, *126*
prevention, 107–108
treatment, 127–128, **127**
in the queen, 239–241
causes, 239–240
and Cesarean section, 241
clinical signs, 240
diagnosis, *240,* 240–241
treatment, 241

ear canals, opening of, in pediatric animals, 277, **278**
eclampsia. *See* hypocalcemia
electrical impedance and breeding management in the bitch, 99
electroejaculation in feline semen collection, 193–194, *194*
en-bloc ovariohysterectomy
in the bitch, 65
in the queen, 208
endoscopes
use in transcervical insemination in the bitch, 55–56
use in vaginoscopy in the bitch, 13, 16
Esterisol™, 89
estradiol cypionate, use in pregnancy termination
in the bitch, **112**
in the queen, **234**
estrogens, use in pregnancy termination
in the bitch, **112**
in the queen, **234**

estrous cycle
in the bitch
normal parameters, 71, *72*, 73
and puppy vaginitis, 134
vaginal cytology, *9*
in the queen
normal parameters, 213–214, *214*
seasonality, 211, 213
estrus
after ovariohysterectomy in the queen, 255–256, *256*
induction protocols
in the bitch, **166**
in the queen, 254
persistent
in the bitch, 161–163, *162*
in the queen, 251–252, *252*
stage in the estrous cycle,
in the bitch, 7, 71, *72*
in the queen, 213, *214*
extenders, use in canine semen preparation, 43, 48
eyelids, opening of, in pediatric animals, 277, **278**

fading. *See* failure to thrive
failure to thrive, in pediatric animals, 293–297, **294,** *295, 296*
false pregnancy. *See* false whelping
false whelping, 73, 83
feline coronavirus, 227
feline infectious peritonitis (FIP), 227
feline leukemia, 227
ferning of vaginal fluid and breeding management in the bitch, 99
fertility improvement and dietary supplements in the bitch, 105–106
finasteride, use in treating canine benign prostatic hypertrophy and hyperplasia, 174
fine-needle aspirate, technique for, canine
prostatic, 37–38
testicular, 39–40
FIP. *See* feline infectious peritonitis
flea control products, safety for use in pediatric animals, 289–290, **289**
fluid therapy in pediatric animals, 296
follicular ovarian cysts,
in the bitch, 161–162, *162*
in the queen, 251–252, *252*
freezing centers, certified by American Kennel Club, 47
frozen semen, preparation of, 47–49

gestational diabetes mellitus and pregnancy loss in the bitch, 147, 148
glucose concentration in vaginal fluid and breeding management in the bitch, 99
GnRH. *See* gonadotropin releasing hormone
gonadectomy. *See* castration, ovariohysterectomy
gonadotropin releasing hormone (GnRH)
use in canine semen collection, 117
use in contraception
in the bitch, 88
in the male dog, 89
use in ovulation induction in the queen, 226

heart rate, in pediatric animals, 277
heat. *See* estrous cycle, estrus
hemacytometer, use in semen evaluation, 30–31, *31*, 196
hereditary conditions, tests for, canine, 43, 47, 51, 55, 57, 91, 92
high-morbidity cancers
and canine castration, 86
and canine ovariohysterectomy, 82
homemade milk replacers for pediatric animals, **286**
hormones, use in contraception
in the bitch, 87–88
in the male dog, 89
in the queen, 225–226
hypocalcemia in the bitch, 129–130
hypoluteoidism and pregnancy loss
in the bitch, 146–148
in the queen, 244
hypo-osmotic swelling test, use in canine semen evaluation, 32
hypothermia in pediatric animals, 279
hypothyroidism, effect on canine fertility, 93–94

immunocontraception
in the bitch, 88
in the male dog, 89
in the queen, 226
inbreeding and small litter size in the bitch, **108**
infertility
in the bitch, 169–171, **170**
in the male cat, 261
in the male dog, 181–183, **182**
in the queen, 259–260, **260**
intraosseous catheter, technique for placement of in pediatric animals, 269–270, *269*
intratesticular injection for contraception, canine, 89
isoerythrolysis, neonatal, in the kitten, 303–304

joint abnormalities in pediatric animals, assessment of, 279
juvenile vaginitis. *See* puppy vaginitis

kittens, resuscitation of, **204,** 205

lack of cycling. *See* persistent anestrus
lactation. *See also* agalactia, hypocalcemia, mastitis
  and depletion of calcium in the bitch, 105, 106, 129
  use of prolactin-inhibiting drugs to inhibit, in the bitch, 73, 83
LH. *See* luteinizing hormone
litter size in the bitch, 107, **108**
lordosis posture in the queen, 213, *214*
Lutalyse™. *See* prostaglandin F2alpha
luteinizing hormone (LH) assays, use in breeding management in the bitch, 97
lymphoid follicles on vaginal mucosa in the bitch, 16, *16*

mammary hypertrophy, feline, 221, *221,* 222
mammary neoplasia and ovariohysterectomy
  in the bitch, 81, 115
  in the queen, 219–220, 235
mastitis
  in the bitch, 157–159, *158*
  in the queen, 249
mechanical barriers for contraception
  in the bitch, 88–89
  in the male dog, 89
megestrol acetate
  usc in thc bitch, 87
  use in the queen, 225
metritis in the bitch, 131
mibolerone
  use in the bitch, 88
  use in the queen, 225–226
mifepristone, use in pregnancy termination in the bitch, **113**
milk replacers for pediatric animals, 285–287, **286**
"mismating" shot, **112,** 161, **234,** 251
morphology, in semen evaluation
  canine, 31–32, *32,* 75, **76**
  feline, 196, 215, **215**
motility, in semen evaluation
  canine, 29, **76**
  feline, 195, **215**
Mycoplasma, 91–92, 95–96

neonatal isoerythrolysis in the kitten, 303–304
neonatal mortality, 305–306
Neutersol™, 89
normal physical examination findings in pediatric animals, 277–279, **278**

obesity
  and canine castration, 85
  and canine ovariohysterectomy, 82
  and feline castration, 223
  and feline ovariohysterectomy, 219
OFA. *See* Orthopedic Foundation for Animals
OHE. *See* ovariohysterectomy
oligozoospermia, canine, 182–183
omega-3 fatty acids, 105
orphan pediatric animals
  dietary recommendations, 285–287, **286**
  environmental temperature recommendations, 285
Orthopedic Foundation for Animals (OFA), and tests for hereditary conditions, canine, 43, 47, 51, 55, 57, 91
otoscopes, use of, 13, *14*
Ovaban™. *See* megestrol acetate
ovarian remnant syndrome in the queen, 255–256, *256*
ovariohysterectomy (OHE)
  in the bitch
    best age for, 81–82
    best time to perform relative to heat, 83–84
    concurrent with Cesarean section, 63, 64, 65, 83
    and cranial cruciate ligament injury, 82
    en-bloc, 65
    and high morbidity cancers, 82
    and mammary neoplasia, 81, 115
    and obesity, 82
    and persistent estrus, 162
    and pyometra, 81, 151–152
    and sexually dimorphic behavior, 81
    and urinary incontinence, 81–82
  in the queen
    best age for, 219–220
    best time to perform relative to heat, 221–222
    concurrent with Cesarean section, 207, 208
    en-bloc, 208
    estrus after, 255–256, *256*
    and mammary neoplasia, 219–220, 235
    and obesity, 219
    and persistent estrus, 252
    and pyometra, 219, 247
ovulation induction for contraception in the queen, 226, 229

oxytocin
and poor mothering behavior, 306
use in treatment of dystocia
in the bitch, 127
in the queen, 241

paraphimosis, recurrent, canine, 179–180
parasites and diarrhea in pediatric animals, 299–300
PennHIP. *See* University of Pennsylvania Hip Improvement Program
persistent anestrus
in the bitch, 165–167, **166**
in the queen, 253–254
persistent estrus
in the bitch, 161–163, *162*
in the queen, 251–252, *252*
persistent male behavior after feline castration, 257–258, *257*
pharmacokinetics of commonly used drugs for pediatric animals, **272,** 273
physical barriers for contraception
in the bitch, 88–89
in the male dog, 89
physical changes in the bitch due to heat, 99
physical examination findings in pediatric animals, normal, 277–279, **278**
physical pre-partum changes in the bitch, 109–110, **110**
physiologic pre-partum changes in the bitch, 109–110, **110**
pipettes, use in vaginal insemination in the bitch, 51–52, *52*
post-estrus stage in the estrous cycle in the queen, 213
prebiotics, use in the bitch, 105, 106
pre-breeding evaluation
in the bitch, 91–92
in the queen, 227
pregnancy
in the bitch
diagnosis, 21, **22–23,** *24*
dietary recommendations, 105, 106, 107, 129
loss, 145–148
physiologic changes, **110**
safety of drugs, 101, **102–103**
termination, 111–114, **112–113**
toxemia, 147
in the queen
diagnosis, 191, **192**
loss, 243–244
safety of drugs, 231, **231–232**
termination, 233, **234**
pre-partum physical/physiologic changes in the bitch, 109–110, **110**
probiotics, use in the bitch, 105, 106
proestrus stage in the estrous cycle
in the bitch, 7, *8–9,* 71, *72,* 73
in the queen, 213
profile, serum chemistry, normal values in pediatric animals, 281, **283,** 284
progesterone
assays, use in breeding management
in the bitch, 97, **98,** 99–100
in the queen, 229–230
concentration, and pregnancy loss in the bitch, 146–148
receptor blockers, use in pregnancy termination
in the bitch, **113**
in the queen, **234**
use in contraception
in the bitch, 87–88
in the queen, 225
prolactin inhibitors
use in lactation inhibition in the bitch, 73, 83
use in pregnancy termination
in the bitch, **112**
in the queen, **234**
Propecia™. *See* finasteride
Proscar™. *See* finasteride
prostaglandin F2alpha
use in canine semen collection, 117
use in pregnancy termination
in the bitch, **112**
in the queen, **234**
use in treating infertility in the male dog, 183
use in treating persistent anestrus
in the bitch, 166
in the queen, 253
use in treating prolonged vulvar discharge after whelping, 132
use in treating pyometra
in the bitch, 152
in the queen, **247**
prostaglandins, use in pregnancy termination
in the bitch, **112**
in the queen, **234**
prostate disease, canine
benign prostatic hypertrophy and hyperplasia, 173–175, *174*
and castration, 85, 174
discoloration of urine, *76*
prostatic biopsy, technique, canine, 37–38
prostatic fine-needle aspirate, technique, canine, 37–38

prostatitis, canine, 177–178, *178*
pseudocyesis. *See* false whelping
puberty, normal age of onset
  canine, 69
  feline, 211
puppies, resuscitation of, 61, **62**
puppy vaginitis, 133–134, *134*
pyometra
  in the bitch, 149–153
    causes, 149
    clinical signs, 149–150
    diagnosis, 150, *150, 151*
    and ovariohysterectomy, 81, 151–152
    treatment, 95, 151–152, **152**
  in the queen, 245–247
    causes, 245
    clinical signs, 245
    diagnosis, 245–246, *246*
    and ovariohysterectomy, 219, 247
    treatment, 247, **247**

queening, normal, 239

radiographs
  abdominal, for pregnancy diagnosis
    in the bitch, **22,** *24*
    in the queen, **192**
  for assessment of bone/joint abnormalities in pediatric animals, 279
  in diagnosis of dystocia
    in the bitch, 126–127, *126*
    in the queen, 240, *240*
  in diagnosis of pyometra
    in the bitch, 150, *150*
    in the queen, 246, *246*
rectal temperature in pediatric animals, 277, 295
recurrent paraphimosis, canine, 179–180
relaxin assay for pregnancy diagnosis
  in the bitch, **23**
  in the queen, **192**
resuscitation
  of kittens, **204,** 205
  of puppies, 61, **62**
retained testes. *See* cryptorchidism

seasonality of estrous cycle in the queen, 211, 213
semen
  canine
    chilled, 43–45
    collection technique, 25, *26,* 27
    evaluation, 29–33, *30, 31, 32*
    frozen, 47–49
    quality, 75–77, **76,** *76,* 181–183
    type, and timing of insemination, 100
  feline
    collection technique, 193–194, *194*
    evaluation, 195–197
    quality, 215–216, **215,** 261
  shipment of, 44, 45, 49
septicemia in pediatric animals, 301–302, *301*
serum chemistry profile, normal values, in pediatric animals, 281, **283,** 284
sexually dimorphic behavior
  and canine castration, 85
  and canine ovariohysterectomy, 81
shipment of semen, 44, 45, 49
side effects of commonly used drugs for pediatric animals, **273**
signs of toxicity of commonly used drugs for pediatric animals, **273,** 274
silent heat, 165
SIPS. *See* subinvolution of placental sites
small litter size in the bitch, 107, **108**
spay. *See* ovariohysterectomy
split heat, 100
standing heat. *See* estrus
sterilization, nonsurgical alternatives
  canine, 87
  feline, 225
stillbirths in pediatric animals, 305–306
subinvolution of placental sites (SIPS), 131–132
surgical intrauterine insemination
  in the bitch, 47, 57–58, *58*
  in the queen, 201

T4. *See* thyroxine
teaser bitch, use in canine semen collection, 25
teeth, eruption of, in pediatric animals, 277, **278**
temperature, and pediatric animals
  environmental, 285, 295–296
  rectal, 277, 295
teratozoospermia, canine, 183
termination of pregnancy
  in the bitch, 111–114, **112–113**
  in the queen, 233, **234**
testicular biopsy, canine, 41–42, *42*
testicular descent, canine, 121–122
testicular fine-needle aspirate, canine, 39–40
testicular neoplasia, canine, 85
testicular ultrasonography, canine, 35–36, *35,* 119
testosterone secretion and persistent male behavior after feline castration, 257–258, *257*
thyroxine, 93

tick control products, safety for use in pediatric animals, 289–290, **289**
tortoiseshell male cat, 261
total number, in semen evaluation
canine, 29–31, *31*, 75–76, **76**
feline, 195–196, **215**, 216
toxemia in the bitch, 147
toxicity, signs of, with commonly used drugs for pediatric animals, **273**, 274
transabdominal ultrasonography for pregnancy diagnosis
in the bitch, **22**, *24*
in the queen, **192**
transcervical insemination in the bitch, 43, 47, 55–56

ultrasonography
canine
in benign prostatic hypertrophy and hyperplasia diagnosis, 173, *174*
and follicular ovarian cysts, 161, *162*
and location of retained testes, 119
of ovaries, in breeding management, 99
in prostatitis, 177, *178*
in pyometra diagnosis, 150, *151*
testicular, 35–36, *35*, 119
transabdominal, for pregnancy diagnosis, **22**, *24*
feline
in pyometra diagnosis, 246
transabdominal, for pregnancy diagnosis, **192**
University of Pennsylvania Hip Improvement Program (PennHIP), and tests for hereditary conditions, canine, 43, 47, 51, 55, 57, 91
Unopette system, use in semen evaluation, 30, *31*, 195–196
urinalysis, normal values, in pediatric animals, 284
urinary incontinence in the bitch and ovariohysterectomy, 81–82
urine sample, techniques for collection, in pediatric animals, 267
uterine disease and pregnancy loss
in the bitch, 147
in the queen, 244
uterine inertia
in the bitch, 125
in the queen, 239

vaginal culture
and antibiotic use during heat in the bitch, 95
in pre-breeding evaluation of the bitch, 91–92
vaginal cytology
in the bitch
interpretation for diagnosis of disease, 7, 10–11
specimen, collection technique, 5–6, *6*
use in breeding management, 7, *8–10*, 97–98, 99
in vaginitis, *134*, 135, *136*
in the queen
interpretation for diagnosis of disease, 190
specimen, collection technique, 187
use in breeding management, 189–190, *189*, *190*
vaginal fluid
ferning of, and breeding management in the bitch, 99
glucose concentration, and breeding management in the bitch, 99
vaginal insemination
in the bitch, 43, 47, 51–54, *52*, *53*
in the queen, 199
vaginal mucosa in the bitch, *15–16*, 16, 98, *98*, 133, 135
vaginal septum in the bitch, *16*
vaginitis
chronic, in the spayed bitch, 135–137, *136*
puppy, 133–134, *134*
vaginography in the bitch, technique, 19–20, *20*
vaginoscopy in the bitch, 13, *14–16*, 16–17, 98
venipuncture, techniques for, in pediatric animals, 265–266, *265*
volume, in semen evaluation
canine, 29, 75, **76**
feline, 195, 215, **215**
vulvar discharge
in the bitch, 10–11
as clinical sign of dystocia, 126
in estrous cycle, 71, *72*
and ovulation, 99
in pregnancy, 143–144
pre-partum, 109
prolonged, after whelping, 131–132
in pyometra, 150, *151*
and vaginitis, 133, 135
in the queen
as clinical sign of dystocia, 240
in pyometra, 245–246, *246*

weaning, 287
weight, in pediatric animals, 277, 293
whelping
abnormal. *See* dystocia
false, 73, 83
normal, 125
prolonged vulvar discharge after, 131–132

zona pellucida (ZP) proteins, for contraception
in the bitch, 88
in the queen, 226
ZP proteins. *See* zona pellucida (ZP) proteins